T0191405

STRUCTURAL OPTIMIZATION METHOD BASED ON SIMULATION

www.royalcollins.com

STRUCTURAL OPTIMIZATION METHOD BASED ON SIMULATION

Mao Huping

Structural Optimization Method Based on Simulation

Mao Huping

First published in 2022 by Royal Collins Publishing Group Inc.
Groupe Publication Royal Collins Inc.
BKM Royalcollins Publishers Private Limited

Headquarters: 550-555 boul. René-Lévesque O Montréal (Québec) H2Z1B1 Canada
India office: 805 Hemkunt House, 8th Floor, Rajendra Place, New Delhi 110 008

ISBN: 978-1-4878-0896-9

To find out more about our publications, please visit www.royalcollins.com.

Brief Introduction

The optimization method of mechanical structures is comprehensively introduced in this book. The book includes the dynamic simulation modeling method and structural dynamic response optimization of spectral element method for different dynamic problems, structural dynamic response optimization considering equivalent static load method of all nodes, and structural static optimization combining elements of optimum methods with structural geometric and engineering requirements. This book is also divided into seven chapters. These chapters are Chapter 1: Introduction; Chapter 2: Structural dynamic analysis method based on spectral element method; Chapter 3: Dynamic response optimization method based on time spectrum element method; Chapter 4: Structural dynamic response optimization method based on the modal superposition method for equivalent static loads of all nodes; Chapter 5: Continuous structural optimization method based on local feature sub-structure method; Chapter 6: Structural dynamic characteristic optimization based on average element energy of the substructure; Chapter 7: A structural dynamic reduction method based on node ritz potential energy and principal degree of freedom.

This book is rich in its content, it covers a new frontier and operability. It can be used as a reference for engineers engaged in the mechanical system or structural analysis and optimization design, and as a reference textbook for graduate or senior undergraduate mechanical structure's optimization courses.

Foreword

Human behavior and natural evolution are closely related to optimization. Engineers adjust various parameters to achieve the best product performance, while manufacturers design different processes to maximize productivity with the lowest production costs or best product performance. The evolution of natural systems is a process of survival of the fittest, which enables various organisms to survive in the worst conditions. Its physical systems will naturally tend to the lowest energy state.

The ultimate goal of product modeling and simulation optimization is to achieve optimal product design. Simulation optimization refers to parameter optimization based on system simulation. It establishes optimization problems for simulation models and uses related optimization search algorithms for solving them. It is an optimization problem based on simulation goals and constraints. Simulation-based optimization is divided into static optimization and dynamic optimization according to the type of simulation problem involved. Simulation-based dynamic optimization can also be divided into simulation-based dynamic response optimization, simulation-based dynamic characteristic optimization, and simulation-based dynamic fatigue optimization. Dynamic optimization based on simulation is not only a problem of simulation optimization but it is also a dynamic optimization problem.

Almost all mechanical structures work in a dynamic load environment, and their various performances are functions that depend on time. To improve the dynamic performance of the machine, dynamic optimization is very necessary. However, it is difficult for the current dynamic optimization design theory and methods to meet the needs of modern product design. Traditional mechanical structure optimization is almost static optimization, which does not consider the dynamic effects due to the dynamic load. That is, under the static force, it is difficult for it to achieve the best machine performance under dynamic load by applying the classic optimization algorithm to optimize. Besides, even if dynamic optimization is performed, it is only directly optimized. Due to the complexity and high time-consuming nature of the dynamic analysis, the direct convergence of dynamic optimization is extremely slow, and the results may even diverge. Given this, in 2007, North University of China sent the author to Huazhong University of Science and Technology to receive a doctor's degree, under the guidance of dynamic simulation optimization experts Professor Chen Liping and Professor Wu Yizhong. The focus of the degree was on the optimization theory and method of the structure under dynamic load. This book systematically summarizes the author's research results starting from 2007.

The chapters of this book are arranged as follows:

Chapter 1: Introduction, which mainly introduces the purpose and significance of the research completed, as well as gives the overview of domestic and foreign research;

Chapter 2: Structural dynamic analysis method based on spectral element method, mainly introduces the application of Chebyshev spectral element method, aggregate element spectral element method in dynamic analysis of structure subjected to impact load and Chebyshev spectrum element method for nonlinear vibration analysis;

Chapter 3: Dynamic response optimization method based on time spectrum element method, mainly introduces the dynamic response optimization model of mechanical structure, dynamic response optimization method, the optimal design of linear single DOF system, the optimal design of linear two DOFs shock absorber and Optimization Design of Dynamic Response of Automotive Suspension System;

Chapter 4: Structural dynamic response optimization method based on the modal superposition method for equivalent static loads of all nodes, mainly introduces the modal superposition method, equivalent static load method, and key time point set;

Chapter 5: Continuous structural optimization method based on local feature sub-structure method, mainly introduces continuous structure optimization problem description and sub-structure methods;

Chapter 6: Structural dynamic characteristic optimization based on average element energy of the substructure, mainly introduces the description of the structural dynamic characteristic optimization problem, structural average element energy, substructure division and implementation of structural dynamic characteristic optimization methods based on substructure average element energy;

Chapter 7: A structural dynamic reduction method based on node ritz potential energy and principal degree of freedom, mainly introduces the calculation of the node Ritz potential energy and the selection of the main degree of freedom. It also explores the construction of the reduction system and the implementation of the structural dynamic reduction method based on the main degree of freedom of the node Ritz potential energy.

The research work involved in this book has been awarded by the National Natural Science Foundation of China project titled, *Method of discrete key points of solution space spectral elements with variable static load and its application in structural dynamic response optimization* (funding number: 51275489) and Shanxi Natural Science Foundation project titled, *Basic research on dynamic response optimization of complex structures based on energy-evaluated PDOFs for multi-substructure reduced equivalent static load and analytical gradient for rigid body mode separation* (funding number: 201701D121082). For this, I would like to express my sincere gratitude.

At the same time, Associate Professor Guo Baoquan of School of Mechanical and Electrical Engineering, North University of China, Energy, and Power of North University of China Associate Professor Zhang Yangang, Associate Professor Wang Qiang, Associate Professor Wang Yanhua, Dr. Wang Jun, Dr. Liu Yong, Dr. Zhao Lihua, Dr. Wang Ying, Dr. Zheng Lifeng, Master Gao Pengfei, and many others, made many valuable suggestions for the study of this book. Master's students Tian Li and Liu Xin have done a lot of work in the writing and revision of this book, and I also sincerely thank everyone!

Because structural optimization is a challenging field of research, and it is still in a vigorous development stage, coupled with the limited level of authors, there are inevitable omissions in this book. Readers are urged to criticize them and help to correct them.

Contents

CHAPTER 1

Introduction

The ultimate goal of product modeling and simulation is to achieve an optimized design of products. Simulation optimization refers to parameter optimization based on system simulation. It is therefore a set of techniques for establishing optimization problems for simulation models and using related optimization search algorithms to solve them. It is also an optimization problem based on simulation goals or constraints. This principle is shown in Figure 1-1, that is, whereby the optimization model is constructed based on the input relationship given by the model simulation, and the best input is obtained by outputting to the optimization algorithm. Many engineering problems such as manufacturing systems, transportation systems, power systems, chemical systems, and others, can be reduced to simulation-based optimization problems. [1–3] Simulation-based optimization is divided into static optimization and dynamic optimization according to the type of simulation problem in question. Simulation-based dynamic optimization can be divided into simulation-based dynamic response optimization, simulation-based dynamic characteristic optimization, and simulation-based dynamic fatigue optimization. Dynamic optimization based on simulation is not only a problem of simulation optimization but also a dynamic optimization problem, too. In a narrow sense, dynamic response optimization is the optimization of the dynamic characteristics of a structure under dynamic loads. Broadly speaking, dynamic response optimization refers to the optimization of objective function optimization or constraint function related to time, while the design variable is independent of time. Of course, the latter also includes the former.

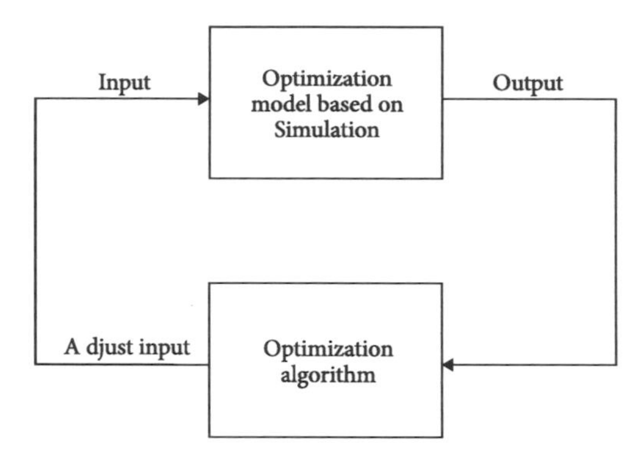

Figure 1-1 Principle of optimization based on the simulation model

The characteristics of simulation-based optimization are as follows:

1) There are two relationships between the input and output of the system. One is the lack of structural information and when there is no analytical expression, which can only be obtained through simulation. The other is that although there is an analytical expression, it is difficult to obtain an analytical solution, such as differential equations or partial differential equations, among others, or use approximate methods to obtain analytical expressions, such as the motion model of the structure.
2) In simulation optimization, one simulation takes less time, the other simulation takes more time, and the lack of simulation optimization algorithms for this situation makes the optimization process very time-consuming and even impossible for optimization.
3) For large continuous structures, optimization is very difficult and is mainly manifested in the fact that the large-scale continuous structure finite element analysis takes a long time, and the efficiency of the optimization process is low. Also, the parameterization of continuous structures is more difficult, except for rod elements, plane elements, beam elements, and others. For simple elements, the element cross-sectional area, element thickness, and beam cross-sectional area are often used as design variables, so it is easy to realize parameterization. Rethinking optimization based on the advantages of the substructure methods and the characteristics of each subfunction in the optimization process is an effective solution.
4) The determination of design variables and their value ranges is the premise of establishing a reasonable optimization model for structural optimization problems. Their values directly determine the convergence and efficiency of optimization problems. Engineers generally take as many design variables as possible based on experience and expand the range of design variables as much as possible, too. This is done to avoid missing key design variables and the important value ranges they

contain, although there is no theory to determine the value range of design variables in this way.

Given the background of structural optimization engineering based on simulation and the above characteristics, it has always been an important subject of common concern for scholars and engineers in many fields, especially machinery manufacturing and aerospace. With the development of computer technology, artificial intelligence technology, and mathematical analysis methods, research on structural optimization based on simulation is now an even more urgent matter.

1.1 The purpose and significance of this book

The research in this book is carried out based on deterministic optimization, and its purpose is as follows:

1) Based on the spectral element method, the Chebyshev temporal spectral element method analysis of vibration problems under arbitrary loads, the dynamic analysis of the aggregate element spectral element method for impact-bearing structures, and the Chebyshev temporal spectral element method analysis of nonlinear vibration will be realized in this title.
2) The structural dynamic response optimization based on the temporal spectral element method will be studied, and the combination of the spectral element method and the modal superposition method will be applied to the structural dynamic response optimization.

   This book further studies the dynamic response design of the system based on the time spectrum element method, discusses the discrete dynamic response in the time domain, converts the differential equations of motion into algebraic equations, and accurately solves the transient response. Using the Guass-Lobbato-Legendre (GLL) point method and the key point method to deal with time constraints, and because of the complexity of transient dynamic analysis and the uncertainty of equivalent static load conversion, an equivalent static load method for all nodes based on modal superposition is also proposed and applied to dynamics Response optimization. Starting from the principle of modal superposition, the relationship between the dynamic response and each mode is further analyzed, and then through the detailed analysis of the principle of the equivalent static load method, the calculation expression of the equivalent static load of all nodes based on the modal response is also given. Then, the key to the equivalent static load method for all nodes in the time point set uses discrete interpolation and differentiation of the spectral elements to obtain the time key points and forms the key time point set with the neighboring GLL points.

3) Continuous structure optimization based on local feature substructure method and structure dynamic characteristic optimization based on substructure average element energy.

Starting from the various sub-functions of the optimization process and the geometric characteristics of the continuous structure, this book divides the continuous structure into parameterized sub-structures, super-units, and state variable sub-structures. It then uses the geometric characteristics of the parameterized sub-structures as design variables to establish continuous structures. The objective function of the evaluation is to take the small continuous structure quality as the optimization goal and take the strain of the continuous structure carried by the parameterized substructure and the state variable substructure as the constraint condition. As such, an optimized mathematical model for continuous structure evaluation implies the whole of the local geometry of the structure is unchanged and does not contain the local structure of the required state variables, that is, the super element). This book uses the gradient-based sequential quadratic programming method to solve the model to achieve a continuous structure efficient optimization method based on local feature substructure. Starting from the establishment of an optimization model, combined with the geometric characteristics of the structure, the overall structure is then divided into multiple quasi-design variable substructures. For the truss structure, each rod is used as a substructure. For continuous structures, multiple elements are used as A substructure, combined with the relationship between the average unit volume strain energy of the structure and the average unit kinetic energy and the contribution of the dynamic response of the structure. It takes the structural unit with the larger average unit volume strain energy as the substructure that should become larger and increases the average unit kinetic energy. The structural unit as a sub-structure whose unit size should become smaller, thereby determines a reasonable range of design variables. This book deduces the relationship between the average unit energy and the dynamic response of the structure, constructs a structural dynamic characteristic optimization model, calls the gradient-based optimizer for iterative optimization, and realizes the optimization of the structural dynamic characteristics based on the average element energy of the substructure.

1.2 Overview of domestic and foreign research

Research on simulation-based structural optimization methods involves both structural optimization and simulation-based optimization.

1.2.1 Overview of structural optimization research

The research content of structural optimization involves a sub-structure method, structural sensitivity calculation, equivalent static load method, spectral element method and structural dynamic response optimization.

1. Sub-structure method

Structural optimization design is based on an accurate structural mathematical model. For large complex structures, the dynamic response is still based on the test results. It is very difficult to study the computer simulation of vibration test. The core problem is how to get an accurate vibration response. Vibration response calculation is much more difficult than static calculation and vibration characteristic calculation. There is an urgent need to improve the calculation accuracy and efficiency of the dynamic response of large complex structures. The substructure method is a suitable choice.

The sub-structure method makes full use of the dynamic characteristics of each subsystem to obtain reliable original system dynamic characteristic parameters or dynamic responses in a simple calculation process. It is an effective calculation method for solving the dynamic analysis of large complex structures. According to different boundary conditions, sub-structure methods can be divided into four categories: [4] constrained sub-structure method, [5] free sub-structure method, [6], mixed sub-structure method, [7] and load sub-structure method. [8] In the early 1960 s, Hurty [9] first proposed the concept of the modal synthesis method and established the basic framework of the constrained modal synthesis method. For more than half a century, the modal synthesis method has made great progress. It has been qualitative and has become a routine method for the dynamic analysis of structures. In 2000, Craig [5] improved it based on Hurty to form a constrained sub-structure method. The fixed interface modal synthesis method refers to additional constraints on all interfaces of the substructure. This method no longer distinguishes between statically determinate and redundant constraint coordinates in the substructure interface, which makes for the fixed interface modal synthesis method. It has there been widely used. The free interface sub-structure modal synthesis method means that the free interface of the sub-structure retains the main modal set as one of the subsets. It also cuts off the connection on the interface, and divides the overall system into several unstructured substructures in space, and then uses the coordination of interface displacement between adjacent components and the equilibrium conditions of the interface. It forces to connect to the previously completely released connection interface into a whole. Because the free interface sub-structure modal synthesis method is more in line with the requirements of the current dynamic test level, this method is very attractive when the reliability of the analytical model needs to be checked through experiments. The free interface sub-structure modal synthesis method is not only better in calculation accuracy than the fixed interface sub-structure modal synthesis method, but also not limited by the number of sub-structure interface coordinates, but this method is also complicated. The hybrid interface substructure modal synthesis method is proposed to overcome the shortcomings of the above two methods. The load sub-structure method improves the accuracy of low-order modes, but as the number of modes increases, its accuracy will also decrease. Based on the above methods, many other methods have been developed and widely used. [10, 11]

The sub-structure method is an advanced finite element analysis method that reduces the calculation time of simulation calculation by processing the solution model in blocks. As long as the block is reasonable and the substructure database of the model is established, the solution efficiency can be greatly improved.

Zhang Yan et al. [12] applied the substructure method to the noise reduction analysis of automobile frames, and the solving efficiency was increased by 94%. Gao Pengfei et al. [13] applied the ANSYS pre-processing module to separate the internal cold oil cavity of the piston and analyzed the mechanical stress and temperature distribution of the piston using the super-unit method in the sub-structure method. The efficiency and accuracy of the structural method applied in the check of the piston structural strength and temperature distribution. Gao Pengfei et al. [14] tried to improve the efficiency of the optimization design of the piston structure, by using a sub-structure method that was first introduced into the optimization design of the structure, and the piston of a supercharged diesel engine was taken as the research object. The research results showed that the optimization iteration convergence time was reduced by 74.14%. In practical problems, it is often necessary to optimize the design of some local features of the model. To overcome the shortcomings of traditional modeling and optimization, such as low efficiency and easy loss of parameterized information, Liu Bo et al. [15] fully used 3D CAD software and The advantages of CAE software are modeled by Pro/ENGINEER software and ANSYS's scripting language APDL as the platform for parametric modeling, and three parametric design methods for local feature optimization are proposed. Li Yun et al. [16] applied the sub-structure method to the large-chassis twin-tower conjoined structure. Through structural static analysis and modal analysis, it fully demonstrated the sub-structure method in the application of large-scale complex model simulation calculation advantage. Chai Guodong et al. [17] applied the sub-structure method in ANSYS to carry out the modal analysis of the electronic equipment box and sought to improve its overall structural rigidity. Zhang Mingming et al. [18] applied the substructure method to the finite element simulation analysis of the diesel engine crankshaft. By dividing the substructure of the crankshaft and establishing the substructure database, the substructure finite element model of the diesel engine crankshaft was combined to carry out the structural static mechanical analysis and modal analysis. Compared with the traditional finite element algorithm, it found that the error of the two calculation methods does not exceed the allowable range of engineering errors, and it once again proved the feasibility of the substructure method in the field of structural finite element simulation calculation. Ding Yang et al. [19] introduced the substructure method into the evaluation of the collapse resistance of the steel frame structure. The evaluation of the collapse resistance of the two five-layer steel frame structures can prove the efficiency and accuracy of this idea. Ding Xiaohong et al. [20] introduced the idea of substructure in the topological optimization design of the car seat frame, and obtained the optimal structure of the seat frame through the stepwise approximation method, reducing the volume of the seat frame. Zhang Sheng et al. [21] also proved that the multiple multilevel substructure method is more efficient and accurate by comparing the accuracy of the multiple multilevel substructure methods and the modal synthesis method in structural modal analysis. Zhang Fan et al. [22] introduced the sub-structure method in the process of topological optimization of the bus body, and designed the part that did not participate in the optimization into a sub-structure, while also connecting the node to the appropriately optimized part. Reducing the matrix order of the overall calculation model improved the efficiency of the body topology optimization design. Li Zhigang et al. [23] divided the substructure of the elevated railway pontoon, and after recombination calculation also proved the high efficiency of the substructure method. The author uses the gradient-based sequential quadratic programming method to solve the model, analyzes and optimizes a diesel engine piston continuous structure as an example, and compares it with the traditional optimization method from the aspects of convergence and efficiency to prove the superiority of this method. [24]

In addition, the author studies the parallel processing and scheduling strategy of the SQP algorithm in multi-domain simulation optimization and proposes an abstract scheduling model in the parallel optimization problem of the SQP algorithm based on multi-domain simulation, that is, the equality constrained discrete variable optimization model, and the algorithm theory. The author also discusses in depth the feasibility of the software, constructs a parallel simulation optimization environment using a cluster system, and implements a parallel optimization module under the independently developed multi-domain unified modeling and simulation platform MWorks. [25]

The dynamic analysis of large complex structures requires the calculation of models with a large number of degrees of freedom. Under the action of high-frequency excitation force, the required calculation step is very small, which leads to an exponential increase in calculation time. To improve the calculation efficiency, the multi-degree-of-freedom model can be replaced with a low-degree-of-freedom model while ensuring a certain accuracy, that is, the model is reduced. The so-called model reduction means that through a certain transformation, the second degree of freedom that has little influence on the overall structural dynamics analysis is expressed by a small number of degrees of freedom that have a large influence on the overall structural dynamic analysis, to reduce the calculation scale. A small number of degrees of freedom are PDOFs. However, how to select PDOFs from huge degrees of freedom is still a very challenging problem in the field of structural dynamics.

However, at present, the academic community has put forward some principles for selecting PDOFs. The representative ones are:

1) Set the vibration direction of the structure as PDOFs.
2) Select the PDOFs at the position where the mass or rotational inertia is relatively large and the rigidity is relatively small.
3) Select the PDOFs at the position where the force or non-zero displacement is applied.

These principles are just guidelines, and when choosing PDOFs specifically, they are more arbitrary. The position and number of PDOFs directly affect the accuracy of modal analysis. Jeong et al. [26] proposed a PDOFs selection method for the damping system based on the degree of freedom energy distribution ratio. To estimate the energy distribution of the structure, the bilateral Lanczos algorithm was used to obtain the Ritz vector, and the obtained Ritz vector was used to calculate the energy distribution matrix. Let the low Rayleigh quotient corresponding to DOFs be PDOFs. Kim et al. [27] proposed a degree of freedom analysis selection method for feature problem reduction. This method selects according to the energy-related modal degrees of freedom of the structural system. The value of the weighted row of the energy distribution matrix is used as an effective method for selecting PDOFs. As a guideline, Cho et al. [28] proposed a unit-level energy estimation method, constructed a small-scale finite element model, calculated the energy of each unit through the Ritz vector, sorted the energy values, and classified the small energy Value as PDOFs.

2. ***Structural sensitivity calculation***

In structural optimization, sensitivity calculation requires a lot of resources, so there are many studies on the efficiency of sensitivity calculation. There are three methods for sensitivity calculation, namely the finite difference method, the analysis method based on discrete equations, and the analysis method based on continuous equations. Among them, analysis methods based on discrete equations are divided into analysis methods and semi-analysis methods, while analysis methods based on continuous equations are considered complete analysis methods. The analysis method includes the direct difference method and the accompanying variable method.

In the finite difference method, the central difference method is commonly used. Among them, the sensitivity of the objective function and the constraint function can be expressed as:

$$\frac{\partial \boldsymbol{g}_j}{\partial \boldsymbol{b}_j}=\frac{\boldsymbol{g}_j\left(\boldsymbol{b}+(\Delta \boldsymbol{b}/2)_i,\boldsymbol{z}\left(\boldsymbol{b}+(\Delta \boldsymbol{b}/2)_i\right),\xi\left(\boldsymbol{b}+(\Delta \boldsymbol{b}/2)_i\right)\right)}{\Delta \boldsymbol{b}_i} - \frac{\boldsymbol{g}_j\left(\boldsymbol{b}-(\Delta \boldsymbol{b}/2)_i,\boldsymbol{z}\left(\boldsymbol{b}-(\Delta \boldsymbol{b}/2)_i\right),\xi\left(\boldsymbol{b}-(\Delta \boldsymbol{b}/2)_i\right)\right)}{\Delta \boldsymbol{b}_i} \tag{1.1}$$

In the formula, $\boldsymbol{b}$ is the design variable vector, ξ is the eigenvalue, and z is the node displacement vector.

Of course, in addition to the center difference method, there are forward difference methods and backward difference methods, but the center difference method has the highest accuracy. The finite difference method is simple to handle and can be obtained by using existing simulation software by treating the simulation model as a black-box function. However, this method is very resource-intensive, especially when the finite element analysis needs to be repeated.

When the simulation code calculation takes little time, the finite difference method is the best sensitivity calculation method. However, for engineering problems, commercial finite element software is generally used for simulation, and the analysis method is more applicable for this at this time. The analysis method based on discrete equations can be expressed as:

$$\frac{\mathrm{d}\boldsymbol{g}_j}{\mathrm{d}\boldsymbol{b}_i}=\frac{\partial \boldsymbol{g}_j}{\partial \boldsymbol{b}_i}+\frac{\partial \boldsymbol{g}_j}{\partial z}\frac{\mathrm{d}z}{\mathrm{d}\boldsymbol{b}_i}+\frac{\partial \boldsymbol{g}_j}{\partial \xi}\frac{\mathrm{d}\xi}{\mathrm{d}\boldsymbol{b}_i} \tag{1.2}$$

In equation (1.2), the most difficult to calculate is $\frac{\partial \boldsymbol{g}_j}{\partial z}\frac{\mathrm{d}z}{\mathrm{d}\boldsymbol{b}_i}$, especially because the calculation of $\frac{\mathrm{d}z}{\mathrm{d}\boldsymbol{b}_i}$ is time-consuming. There are two methods for calculating $\frac{\mathrm{d}z}{\mathrm{d}\boldsymbol{b}_i}$, namely the direct difference method and the accompanying variable method. The latter introduces a companion equation $\boldsymbol{K}\boldsymbol{\lambda}_j=\left(\frac{\partial \boldsymbol{g}_j}{\partial z}\right)^{\mathrm{T}}$ ($j = 1, 2, \cdots, m$), which is

$$\frac{\partial \boldsymbol{g}_j}{\partial z}\frac{\mathrm{d}z}{\mathrm{d}\boldsymbol{b}_i}=\boldsymbol{\lambda}_j^{\mathrm{T}}\left(-\frac{\partial \boldsymbol{K}}{\partial \boldsymbol{b}_i}\boldsymbol{z}^{(k)}+\frac{\partial \boldsymbol{f}}{\partial \boldsymbol{b}_i}\right) \tag{1.3}$$

In the optimization process, the number of times that the accompanying variable method needs to be solved is equal to the number of activated constraints and the number of times that the direct difference method is solved is equal to the number of design variables. Therefore, judging the validity of the two requires a concrete analysis of the problem.

In the analysis method, the difference between the mass matrix and the stiffness matrix is the most difficult. The calculation of mass matrix and stiffness matrix in commercial software usually uses the finite difference method based on finite element, but its accuracy depends on the size of the disturbance size, especially in the shape optimization of large complex structures. The analysis method based on the continuous equation starts from the integral formula and is expressed as

$$\delta\psi = \int_{\Omega} \left(\boldsymbol{g}_z \delta z + \boldsymbol{g}_b \delta \boldsymbol{b} \right) \mathrm{d}\Omega \tag{1.4}$$

In the formula, $\boldsymbol{g}_z = \frac{\partial g}{\partial z}$, $\boldsymbol{g}_b = \frac{\partial g}{\partial \boldsymbol{b}}$, $\delta z = \frac{\mathrm{d}z}{\mathrm{d}\boldsymbol{b}} \delta \boldsymbol{b}$ calculating δz is crucial. Analytical methods based on continuous equations are also divided into direct difference methods and accompanying variable methods.

Since the semi-analytical sensitivity analysis (SAM) method combines the accuracy of the analysis method with the efficiency of the finite difference method and it is suitable for application in commercial software, this method has always been a research hotspot. In 1973, Zienkiewicz and Campbel [29] proposed a semi-analytical sensitivity analysis method. Later, Barthelemy et al. [30] and Pauli [31] found in the research that the semi-analytical sensitivity analysis method has appeared inaccurate in some applications. To solve this problem, Olhoff et al. [32] proposed an intermediate difference scheme to solve the differentiation of the stiffness matrix. In 1993, Cheng and Olhoff's [33] research discovered the real reason for the inaccuracy of the semi-analytical sensitivity analysis method, that is, when there is rigid body motion in the unit, will show unreliable accuracy. The essence is that the rigid body motion is directly related to the truncation error. For this reason, Keulen and Boer [34] proposed a sophisticated semi-analytical sensitivity analysis (RSAM) method, which eliminates sensitivity errors caused by rigid body modes based on accurate rigid body mode differences. However, when the disturbance size is relatively large, the RSAM method cannot obtain sufficient accuracy. Therefore, in the semi-analytical method, the higher-order terms need to be considered. In the expansion of the higher-order terms, the inverse matrix can be expanded by the Neumann series, [35] and the RSAM based on modal decomposition is also studied and described its application in nonlinear structural analysis. Cho and Kin [36] combined modal decomposition and Neumann series to develop the RSAM method.

3. *Equivalent static load method*

The equivalent static load method equates the structural displacement field under the equivalent static load with the displacement field at a certain moment under the dynamic load. [37] The researchers carried out the concept of equivalent static load in the previous study. The concept of equivalent static load method has been extended and it showed that the equivalent static load should not only replace the displacement field generated by the dynamic load but also replace the volume strain energy, that is to say, its replacement effect must include the displacement field and volume strain energy. Driven by the idea of equivalent load, two types of methods have evolved: a) Optimization of the dynamic response of the structure based on the equivalent static load of time-critical points; b) It must be based on the dynamic analysis of all structural dynamic time steps or the designation of the time subdivision points specified by the designer. This also concerns structural dynamic response optimization for static loads. The latter considers all possibilities, and each step of the dynamic analysis of the structure or each point specified by the designer at the time subdivision point is equivalent to a set of static loads. For small load steps, this method takes too long, and for slightly larger load steps, the relative accuracy of the structural dynamic response analysis will be, to some extent affected. Of course, if there is a sufficiently good computing environment, this problem has little effect. If the time subdivision point specified by the designer is adopted, although there would be no accuracy problems, there would be problems such as subdivision. Either way, when considering all the time points, there are huge constraints and load states, which bring great challenges to the optimization algorithm. The author has studied how to efficiently identify key time points. [38] At the key point moments, the dynamic load is more reasonably converted into static load through the volume strain energy equivalent, [39] and then the SQP multi-initial point method can be used to solve it to make it better. In terms of convergence, however, in this study, the load position vector was not considered. This means that the equivalent static load acting position was not considered, but the equivalent static load was applied to the dynamic load acting position or the position where the load should be applied according to previous experience. There is no doubt that there is a certain randomness. Because of the different position of the equivalent static load and the uncertainty of its range of values, the calculation time and value are obviously different. And the essence of solving the equivalent static load is also an optimization problem. In this way, the optimization using the equivalent static load as the design variable needs to determine its value space, which in turn has certain randomness. This randomness will lead to uncertainty in the results, and it is time-consuming to solve the equivalent static load. Aiming at the complexity of the dynamic analysis of the structure and the uncertainty of the conversion of the equivalent static load, the author proposes the equivalent static load method of all nodes based on the superposition of the mold. The author also proposes to apply it to the dynamic response optimization. The results of dynamic response size optimization and mixed size and shape optimization design of the 18-bar truss structure show that this method is feasible and effective. [40] Aiming at the complexity and time-consuming problems of dynamic analysis in structural dynamic response optimization, the author proposes a key time point recognition method based on global dynamic stress solution space spectral element interpolation, which finds the most dangerous moment under the dynamic response of the structure. Specifically, first, the author proposes the use of the modal superposition method to obtain the modal stress distribution of the structure and to calculate the global dynamic stress solution space. Then, to use the spectral element discrete dynamic stress absolute maximum point curve, then use Lagrange interpolation and to call the area fine. The sub-global optimization solver should find the maximum and minimum values of the global dynamic stress, that is, the key time point. [41]

4. *Spectral element method*

The Spectral Element Method (SEM) is based on the weak form of the elastic mechanics' equations and it performs spectral expansion on finite elements. This method has the toughness of the finite element method to adapt to any complex medium model and the accuracy of the spectral method. It is also called a high-order domain decomposition of the finite element method or spectral method.

Patera proposed the spectral element method [42] in 1984 and applied it to fluid dynamics, which combined the flexibility of the finite element method with the processing boundary and structure, as well as the fast convergence of the spectral method. In the case of the same precision, the spectral element method uses fewer units, reducing the computational overhead. The spectral element method includes the spatial-spectral element method, temporal spectral element method, and space-temporal spectral element method. The spatial-spectral element method uses region embedding technology to embed complex geometric regions in practical problems into a regular rectangular region in order to construct an appropriate spectral element space (equivalent to the finite element space of the finite element method), which solves the requirements of the spectral method on the region. [43] The law of time spectral elements constructs the spectral time unit based on finite element space, then interpolates it in each unit, and finally solves the system of linear equations. The space-temporal spectral element method discretizes space or time into grid points corresponding to the zero points of the GLL polynomial or the Chebyshev polynomial in one spectral unit. Lagrange interpolation is performed on these points. [44] In theory, it can interpolate at a certain number of points. When these points are the zero points of the corresponding orthogonal polynomials, the highest interpolation accuracy is obtained. [45]

A lot of progress has been made in many works on the spectral element method. The spectral element method is widely used in the numerical simulation of compressible and incompressible fluids. [46] Pathria [47] uses the spectral element method to solve the elliptic problem in the non-smooth domain. Hesthaven [48] proposed a spectral method using open boundary conditional area decomposition. M. H. Kurdi [44] uses the time spectrum element method as the overall solution for ordinary differential equations. The author [45] uses the time spectrum element method for the dynamic response simulation of structures based on the work of M. H. Kurdi. When it comes to wave propagation [49], in order to expand the adaptability of the spectral element method and for the structural dynamic response to impact load, starting from the discrete scheme of the spectral element and according to the characteristics of the impact load, the size of the spectral element is centered on the maximum point of the impact load. This is done by expanding it to both sides by a certain proportion in equal proportions to achieve the unit size and load characteristics. On this basis, the dynamic equations are converted into first-order linear differential equations. The discrete linear equations are obtained by the Bubnov-Galerkin method and solved by Gaussian elimination. Compared with the isometric spectral element method, this can prove the feasibility and effectiveness of the method. [50] The author studied Chebyshev temporal spectral element method to solve the vibration problem under arbitrary load. Starting from the Bubnov-Galerkin method, the Lagrange interpolation of the center of gravity at the pole of the second type of Chebyshev orthogonal polynomial was used to construct the node basis function and analyze its characteristics. Galerkin spectral element discrete scheme for vibration problems under arbitrary loads uses the least squares method to solve linear equations. Similarly, it can use the vibration problems under linear load, triangular load, half-sine wave load, and cantilever beam vibration under sine load as examples. The feasibility of this method was verified and compared with the collocation method, which further explained the high accuracy and reliability of this method [51]. The author also studied the Chebyshev temporal spectral element method to solve the nonlinear vibration problem. Starting from the Bubnov-Galerkin method, the center of gravity Lagrange interpolation at the pole of the second type of Chebyshev orthogonal polynomials was used to construct the node basis function to analyze its characteristics. The nonlinearity was derived Galerkin spectral element discrete scheme for vibration problems, using the Newton-Raphson method to solve nonlinear equations. Much the same, for nonlinear simple pendulums, it is also necessary to combine the dichotomy method and the center of gravity Lagrange interpolation to solve the angular frequency. The use of the Duffing type nonlinear vibration and nonlinearity, the simple pendulum vibration problem (as an example) shows the feasibility and high accuracy of this method. [52] Aiming at the large degree of freedom of structural dynamic response equations, and the spectral element method is the product of the matrix of large degrees of freedom and the overall time matrix tensor. It requires a large memory and takes a long time to solve. The author proposes a stepwise temporal spectral element method. The author proposes to divide the simulation time into very small periods, then divide the units in each time period, and use spectral expansion approximation in each unit so that the processing has the flexibility of finite elements to deal with complex structures and boundaries, as well as the high precision of the spectral method. It would have rapid convergence, the efficiency of gradually dividing the simulation time, and other characteristics. [53] Given the shortcomings of traditional structural dynamic response optimization methods, the author proposes a combination of meta-model hybrid adaptive optimization and temporal spectral element method. Starting from the Bubnov-Galerkin method, the discrete dynamic response in the time domain is discussed in-depth, and the overall structural dynamic equation transforms into algebraic equations to solve the dynamic response accurately and efficiently. According to the characteristics of the optimization problem, it can adopt the adaptive strategy to select the corresponding meta-model for optimization. In the optimization process, it uses a uniform grid to obtain the number of potential points, and localizes the fusion of optimization and multi-model hybrid adaptive methods to make the optimization results more reliable. To deal with time-related constraints, the author proposes a key point set method that combines key points and their neighboring Gauss-Lobatto-Legendre points into a single set. But whether it is accuracy or efficiency, the combination of meta-model hybrid adaptive optimization and temporal spectral element method is better than the key point set method. [54]

The accuracy of the spectral element method can be achieved either by increasing the degree of freedom of the spectral method on each unit or by increasing the number of units. The best case is that the degrees of freedom on each unit can be adjusted freely without restricting each other. Only this kind of spectral element method has enough flexibility.

In general, it is not appropriate to use the temporal spectral element method to solve structural dynamic equations, because if there are too many discrete elements, the process of solving linear equations takes time and affects its application. Also, if there are too few discrete elements, the dynamic response is not accurate enough to meet the engineering requirements. In addition, because the finite element of the structure is discrete, the degree of freedom is generally very large, and the time spectrum element method to solve the structural dynamic equation requires matrix inversion operation, making the engineering application is difficult. The method to solve this problem is to use a time-segment solution, and to adopt the element-by-element technology in each segment. [55] However, this cannot fundamentally solve its engineering application problems. In the analysis of engineering structural dynamics, the number of structural elements is relatively large, and the matrix inversion operation of each element by element-by-element technique also hinders its engineering application.

*5. **Structural dynamic response optimization***

In the 1960 s, Niordson [56] put forward the concept of structural dynamic characteristics optimization and conducted corresponding research, and then opened the prelude of structural dynamic optimization-structural dynamic characteristics optimization. The early structural dynamic characteristic optimization method is a distributed parameter structure optimization method, which belongs to the analytical method. Due to the difficulty of solving partial differential equations in this method, it is only suitable for some simple structures, and it is useless for large and complex structures. Subsequently, the criterion method and the mathematical programming method have been developed, and this part of research is now relatively mature.

In structural design, it is critical to obtain external loads accurately, but it is difficult in most situations. Therefore, a static load is generally set for static optimization design. Strictly speaking, the load on the structure is dynamic. The dynamic factor method can transform the dynamic optimization problem into a static optimization problem, but such processing often results in over- or under-design of the structure, so it is more reasonable to directly use dynamic loads for dynamic optimization of the structure. [57]

Wang et al. [58] applied a mathematical programming method to optimize the design of the plane orthogonal steel frame structure under dynamic load. During optimization, the natural frequency of the structure is not less than a certain value, the maximum dynamic displacement and dynamic stress are not more than a certain value of constraint conditions, with the total mass of the structure as the optimization goal. Yet, it also does not consider the structural damping. In addition, its research found that the feasible region of structural parameters is generally discontinuous when it is designed for dynamic response optimization. Qin Jianjian et al. [59] used the finite element structure simulation analysis method based on ANSYS to establish the finite element model of the diesel engine connecting rod based on the ANDL language. Based on the finite element analysis, the use of ISIGHT integrated optimization software combined with a multi-island genetic algorithm to optimize the design of the connecting rod shaft, so that the quality of the shaft is reduced by 6.02%. Lin et al. [60] used the element reconstruction method and the progressive algorithm of the shape of the structure. The structural dimensions and node coordinates of the constraints and dynamic constraints are effectively optimized. Pantelides et al. [61] aimed at its shortcomings that the initial design point was not feasible at the time of optimization, and the general optimization method may not converge to the global optimal solution. The MISA algorithm (improved simulated annealing method) was applied to simultaneously consider the structure. The dynamic optimization problem of dynamic displacement and dynamic stress constraints and the MISA algorithm were compared with the general optimization method to verify the advantages of the MISA algorithm. Min et al. [62] used homogenization and direct integration methods to carry out topological optimization design of the thin plate structure under the impact load. This work was forward-thinking and pioneering. Du et al. [63] were different from the previous dynamic optimization studies that considered natural frequency and dynamic response displacement. They mainly considered how to reduce the sound radiation intensity of the structure. In the analysis, the coupling effect of the structure and the sound propagation medium was ignored. Under the action, the sensitivity of its structural parameters was calculated and analyzed. On this basis, the topology optimization of the vibration structure was successfully carried out.

In summary, the structural dynamic response optimization design is divided into three research directions: a) the treatment of time-related constraints; b) sensitivity analysis; c) approximation. The research group where the author was working had studied the two directions of a) and c), and proposed a method of processing time-dependent constraints based on GLL point sets. [45, 63, 64] This method uses fewer spectral elements that meet the accuracy requirements to solve the differential equations of motion. Within each element, it performs a one-dimensional search for its high-order Lagrange interpolation function to find the absolute value extreme point of the element. Two adjacent GLL points are used as constraints, and other points near the maximum value are included to form a GLL point set constraint. When the displacement of the component under the dynamic load is very small or only the displacement in a certain direction is considered, its geometry or size will more or less change. At this time, each unit inside the component will have a shape caused by the dynamic load. In terms of relative change, the author's research group corresponds the dynamic load change to the volume strain and proposes the equivalent volume strain static load method. [65] The functional relationship between the volume strain and the dynamic load change under the static load is derived to achieve the equivalent volume strain.

1.2.2 Overview of Simulation-based Optimization Research

During the simulation optimization iteration process, a simulation program needs to be called to calculate the values of the objective function and the constraint function. The response surface method is an effective way to improve simulation optimization efficiency. Response surface refers to the functional relationship between the output response variable and a set of input variables (x_1, x_2, ···, x_n). Generally, the response surface reflects an approximate model of a complex and computationally intensive original model (such as multi-domain simulation model, FEA model, CFD model, etc.). Therefore, the response surface is also called the proxy model (Surrogate) or meta-model, that is, the model of the model.

The response surface method (RSM) refers to the approximate method of solving the design or analysis of the original model by constructing the response surface of the original model. According to reports, [66] Ford Motor Company needs 36 to

160 hours to perform simulation analysis of a car crash model. To achieve the design optimization of the two variables of the model, assuming that on average it takes 50 iterations for optimization, and each iteration requires a simulation calculation, it takes 75 days to 11 months to obtain the solution to the optimization problem. Similarly, it may take longer to realize the design optimization of the FEA model or CFD model. This is almost unacceptable in practice. Therefore, in the past 20 years, the response surface method came into being and has developed rapidly. This method can reduce the number of simulations of the original model in the optimization iteration process, and the response surface is constructed based on the sampling point data, and the sampling point estimation calculations are independent of each other (the traditional optimization iteration process is sequence estimation). It is easily obtained through parallel computing. Therefore, this method can greatly improve the design optimization efficiency of complex analysis models.

According to the literature, [67, 68] the role of RSM includes the following four aspects:

1) Model approximation: This is the basic function of RSM. An approximate model of the response surface of a complex original model in its global definition domain can be established, and the approximate model can be used to realize rapid estimation of new unknown design points.
2) Exploration of design space: Based on the established response surface model, it can help engineers or designers perform parameter experiments, sensitivity analysis, and visualization of the functional relationship between response variables and input parameters, thereby helping engineers to better understand the characteristics of the original model.
3) Accurate expression of optimization problems: Design space exploration based on response surface models, especially sensitivity analysis, can help designers construct more accurate design optimization problems. For example, those non-sensitive parameters can be eliminated from the design variable set, thereby reducing the dimension of the design variables. According to the parameter experiment, the search interval can also be reduced, thereby reducing the sampling interval, and thus reducing the number of optimization selections. Similarly, through analysis, a multi-objective design optimization problem may be simplified into a single-objective optimization problem. But, with a simple single-objective design optimization problem, through exploration of the design space, it may be necessary to establish a multi-objective design optimization problem to solve.
4) Support for optimization methods: This is currently the main application area of RSM. Using the established response surface model can assist in completing various design optimization problems involving original model simulation, such as global

optimization, multi-domain simulation optimization, multi-objective optimization, multi-disciplinary design optimization, and probabilistic design optimization, including reliability optimization, robust optimization, and others.

An important part of the response surface method is to construct the response surface model. There are many types of response surface models, which are suitable for different needs. Commonly used are polynomial regression surrogate (PRS, usually called RSM model), Kriging interpolation model, radial basis functions (RBF) model, support vector regression (SVR) model, neural network (NN) model, and RBNN model based on the mixture of RBF and NN; There are also multivariate adaptive regression spline (MARS) based on spline Model, BMARS (B-spline MARS) model, and NURBS model, as well as inductive learning model, least interpolating polynomial (LIP) model, etc.

To construct a response surface model, the first step is to sample in the design space to form the design point set S. Then, it is to perform simulation calculations (also called "Expensive Calculation") to obtain the response data set Y; finally, according to different algorithms S and Y construct different response surface models. The experimental design method provides a variety of sampling strategies, mainly including two categories: edge distribution and space-filling. The edge distribution type is also called the classic sampling method. With this method, the sampling points are mainly distributed near the boundary of the design domain. Typical edge distribution sampling methods are full/fractional factorial design, center composite design, Box-Behnken, etc., as well as Taguchi, D-Optimal Plackett-Burman, and other methods. The full spatial distribution type refers to the sampling points covering the entire design domain. The sampling methods of this type include simple grid, Latin hypercube design, orthogonal alray, random sampling, uniform design, scrambled nets, Monte Carlo simulation, and Hammersley sequence design, etc.

In general, it is not appropriate to construct a response surface model by sampling it once, because if there are too many sampling points, the construction process is time-consuming and may affect the use. If there are too few sampling points, the constructed response surface model is not accurate enough to be difficult for the Meet application needs. In addition, because the behavior of the original model is unknown, it is difficult to determine a suitable sampling method. The method to solve this problem is to construct a sequence response surface model based on sequence adaptive sampling. The main idea of sequence adaptive sampling is to determine the density of sampling points according to the error between the approximate value and the true value. The sequential exploratory experiment design (SEED) [69] method is representative of this type of experiment design method. The simulated annealing algorithm is used in i-Sight software for adaptive sampling.

When using the response surface method to solve actual engineering problems, the following five factors must be comprehensively considered:

1) The accuracy of the response surface. There is no doubt that the accuracy of the response surface model is a basic requirement for approximation.
2) The number of simulation estimates of the original model, that is, the total number of sampling points needed to construct the response surface model. Due to the high computational cost of each estimate, it is necessary to limit the number of simulations of the original model. Under the same accuracy, the less the number of simulation estimates of the original model, the higher the efficiency of the construction of the response surface model.
3) Time to construct and optimize response surface. As mentioned earlier, the construction of response surface models is often a step-by-step process. The sequential adaptive sampling construction of step-by-step response surface models is a research hotspot of current response surface methods, especially when the number of sampling points gradually increases. How to quickly update the response surface model to implement its incremental construction algorithm is a subject worth exploring for various response surface methods.
4) The storage space occupied by the response surface model. Obviously, the larger the memory space occupied by the response surface model itself, the slower the construction process and the slower the evaluation using it. Therefore, under the same circumstances, the less information contained in the response surface model itself, that is, the smaller the memory space occupied, the better.
5) Use the response surface model to estimate the speed of a given point. The ultimate goal of building a response surface model is to use it for valuation, and usually, this valuation process is called "cheap calculation," therefore, it is heavily executed in applications such as optimization iterations. It can be seen that if the evaluation speed of a given point is too slow, it will make the ideal "cheap calculation" not "cheap."

1.3 The main research content, main innovations, and organizational structure of this book

1.3.1 Main research content

This book focuses on practical engineering problems based on the optimization of the structure of the simulated black-box function model and the effective studies and efficient optimization techniques. The main research points are as follows:

1) This book studies Chebyshev temporal spectral element method to solve the vibration problem under arbitrary load. Starting from the Bubnov-Galerkin method, it deeply analyzes the node basis function constructed by the center of gravity Lagrange interpolation at the pole of the second type of Chebyshev orthogonal polynomial and its characteristics. It also looks at the derived Galerkin spectral element discrete scheme for vibration problems under arbitrary loads, and the least square method that is used to solve the linear equations. To expand the adaptability of the spectral element method, for the structural dynamic problem under impact load, starting from the discrete scheme of the spectral element and according to the characteristics of the impact load, the size of the spectral element is centered on the maximum point of the impact load at a fixed ratio, among other factors. The two sides are enlarged to achieve the unit size and load characteristics. On this basis, the dynamic equations are converted into first-order linear differential equations, the discrete linear equations are obtained by the Bubnov-Galerkin method, and solved by Gaussian elimination. The study uses the Chebyshev temporal spectrum element to solve the nonlinear vibration problem. Starting from the Bubnov-Galerkin method, the nodal basis functions constructed by the Lagrange interpolation center of gravity at the poles of the second type of Chebyshev orthogonal polynomials and their characteristics are also deduced. The Galerkin spectral element discretization scheme for linear vibration problems and the Newton-Rapnson method are used to solve nonlinear equations.
2) This book studies the system dynamic response design based on the time spectrum element method. It also discusses in-depth the discrete dynamic response in the time domain, the transformation of the differential equations of motion into algebraic equations, accurately solving the transient response, and using the GLL point method and key point method to deal with time constraints. Taking the design of the spring shock absorber as an example, introducing artificial design variables, detailed analysis of the advantages and disadvantages of the two processing constraints, it also shows the correctness of this method. These points can provide a reference for further research on dynamic response optimization, such as studying the sensitivity analysis of complex systems on this basis to improve the practicality of this method.
3) Given the complexity of transient dynamic analysis and the uncertainty of the conversion of equivalent static load, this paper proposes the equivalent static load method of all nodes based on modal superposition and applies it to the optimization of dynamic response. Firstly, based on the principle of modal superposition, the relationship between the dynamic response and each mode is analyzed. Then, through detailed analysis of the principle of the equivalent static load method, the calculation expression of the equivalent static load of all nodes using the modal response is given. Finally, the equivalent static load method of all nodes in the key

time point set is proposed. The time key points are obtained by discrete interpolation and differentiation of the spectral elements, and the key time point sets are formed with the neighboring GLL points.

4) In order to achieve the feasibility and efficiency of a continuous structure optimization, this book proposes an optimization method based on local feature substructure. Starting from the analysis of the various sub-functions of the optimization process and the geometric characteristics of the continuous structure, the continuous structure is divided into parameterized sub-structures, super elements, and state variable sub-structures, and the geometric characteristics of the parameterized sub-structures are used as design variables to establish continuous structure evaluation. The function of the objective is to take the minimum continuous structure quality as the optimization goal and use the continuous structure stress and strain carried by the parameterized substructure and the state variable substructure as the constraint conditions to establish the optimal mathematical model for continuous structure evaluation, which contains a super elements' sub-structure. The local geometry of the structure is unchanged and does not contain the local construction of the required state variables. For the model solution, the gradient-based sequential quadratic programming method is used for the solution. Taking a diesel engine piston continuous structure optimization as an example for analysis and optimization, and compared with traditional optimization methods from the aspects of convergence and efficiency, it shows the rationality and superiority of the method in this book.
5) To improve the feasibility and efficiency of structural optimization, this book proposes a structural dynamic characteristic optimization method based on the average element energy of substructures. Starting from the establishment of an optimization model, the overall structure is divided into multiple sub-structures of quasi-design variables in combination with the geometric characteristics of the structure. For the truss structure, each rod is used as a sub-structure; for continuous structures, multiple elements are used as a substructure, combined with the relationship between the average element volume strain energy and the average element kinetic energy of the structure and the dynamic response contribution of the structure, while the structural element with larger average element volume strain energy is regarded as the sub-structure with the larger size, and the structural element with larger average element kinetic energy is regarded as the sub-structure with smaller element size. This is done to determine the reasonable range of design variables. This book deduces the relationship between the average element energy and the dynamic response of the structure, constructs an optimization model for the dynamic characteristics of the structure, and calls the gradient-based optimizer for iterative optimization.
6) To improve the efficiency of structural dynamic analysis, this book proposes a method for the dynamic reduction of structures based on the master degrees of freedom of

node Ritz potential energy. The principle of the improved reduced system method is described, while the process of extracting the Ritz vector is analyzed, the node Ritz potential energy is defined, and it is used as the basis to capture the master degrees of freedom that can accurately reflect the dynamic characteristics of the structure, and the node Ritz potential energy is given. In regards to the calculation formula, the calculation and analysis of two examples of cylindrical curved plates and crankshafts, the feasibility and superiority of the method of this book are also verified. The results of the study show that it is easier to capture the dynamic characteristics of the accuracy by defining the node Ritz potential energy in the Ritz vector space. The weighting coefficient can improve the accuracy of the higher-order frequency. In the structure reduction, the main degree of freedom is about 1/3 of the total degree of freedom is most appropriate.

1.3.2 Main innovations

The main innovations studied in this book are as follows:

1) Applying the temporal spectral element method to the simulation of mechanical dynamic problems, the Chebyshev spectral element method for vibration analysis of arbitrary loads, the aggregate element spectral element method for dynamic analysis of structures subjected to impact loads, and the Chebyshev spectral element for nonlinear vibration analysis are proposed. In terms of the method, the book applies the temporal spectral element method to the dynamic response optimization of mechanical systems. The spectral element method is used to solve the dynamic response of the mechanical system, which improves the shortcomings of the large error in the traditional solution of the dynamic response and achieves the spectral convergence accuracy. In this way, the dynamic response optimization can find the objective function that satisfies all the time constraint changes on the hypercurve or hypersurface. In dealing with constraints, two methods of GLL point method and key point method are used, and the advantages and disadvantages of these two methods are compared.
2) Applying the sub-structure method to continuous structure optimization. The three factors of optimization are combined with the geometric characteristics of the continuous structure, and the parameterized substructure, super-element, and stated variable substructure are defined respectively. The design variables correspond to the parameters of the parameterized substructure, the objective function and the constraint function correspond to the response values of the state variable substructure. The super-element is a part of the continuous structure that contains neither design variables nor state variables. This definition can not only

improve the optimization efficiency but also reduce the difficulty of finite element parameterization.

3) A method for optimizing the dynamic characteristics of structures based on the average element energy of the sub-structure is proposed. For some special structures, such as truss structures, the structure is divided into multiple quasi-design variable sub-structure, and the enlarged element sub-structure, and the reduced sub-structure. These are determined by the average element volume strain energy and kinetic energy, respectively, and then determine the range of design variables, which lays the foundation for establishing an accurate optimization model.

1.3.3 Organizational structure

The organizational structure of each chapter is shown in Figure 1-2.

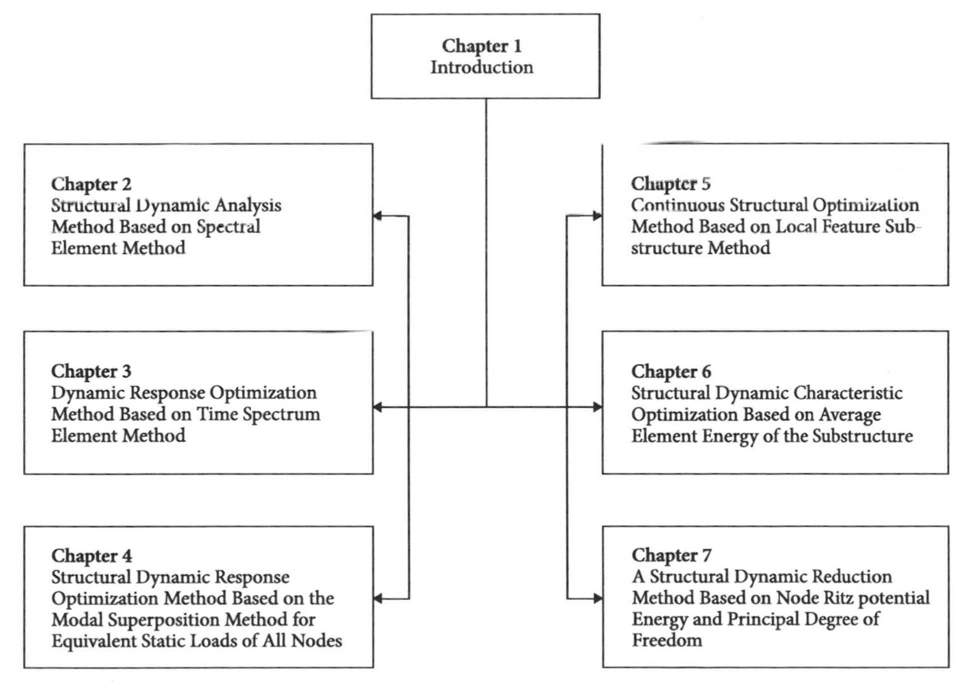

Figure 1-2 Organization structure of each chapter

CHAPTER 2

Structural Dynamic Analysis Method Based on Spectral Element Method

Vibration analysis plays an irreplaceable role in transient analysis, mechanical engineering, fault diagnosis [70] and other fields, and is an important analysis in mechanical design. [71] How to obtain accurate vibration response has always been the focus of researchers. [72]

Numerical methods for vibration analysis are developing constantly, and each method has its advantages and disadvantages. Among them, the representative ones are perturbation transfer matrix method [73] and finite difference method (FDM). The former is to consider the randomness of the system, while the latter is to directly transform differential equations (groups) into algebraic equations, whose mathematical concepts are simple and intuitive, expressions are simple, and programming is easy. The time step directly determines the convergence and accuracy of the calculation [45]. The research on Chebyshev pseudo-spectral method [74–77] shows that the simulation time range is small and high accuracy can be obtained; otherwise, the obtained solution would be meaningless. By using the variational principle of finite element method and the advantages of difference method, the approximate solution with a certain accuracy can be further obtained, and when the shape function is also the interpolation basis function of the zero or pole of orthogonal polynomial, it is called the spectral element method. At the same time, when the orthogonal polynomial is Chebyshev orthogonal polynomial, it is called the Chebyshev spectral element method.

Steven Orszag proposed the spectral method in 1969. [42, 78, 79] A large number of studies have shown that it has the advantage of rapid convergence of higher-order numerical analysis, but its development is limited by its inability to deal with complex spatial domain and other disadvantages. Considering the flexibility of the low-order finite element method in the unstructured domain and the high accuracy and convergence characteristics of the spectral method, Patera [42] proposed the spectral element method in 1984, adopted the Lagrange interpolation and P-type node basis function in GLL, and applied it to fluid dynamics. Dimitri Komatitsch [80] combined the flexibility of the finite element method with the accuracy of the spectral method, and introduced it into the three-dimensional seismic wave calculation. The wavefield on the element was discretized by using the higher-order Lagrange interpolation, and then the element was integrated according to Gauss-Loba-Legend integral rule. In recent years, spectral element method has been applied in many fields of science and engineering. [44, 81] Legendary quadrilateral spectral elements are used to approximate the Black-Scholes equation and apply it to the pricing of European rainbow and basket options. [82] The shallow water equation in spherical geometry was analyzed by spectral element method and compared it with other models [83] by using Lobatto-Legendre orthogonal polynomials to expand the time-domain spectrum of the vibration differential equation, the authors [64] proposed a GLL point set method to deal with time-dependent constraints by using the Galerkin spectral discrete scheme to obtain precise solutions and further proposed a stepwise time spectral method to reduce CPU time. P. Z. Bar-Yoseph et al. [84] used the temporal spectral element method to solve nonlinear chaotic dynamic systems. All the while, U. Zrahia and P. Z. Bar-Yoseph [85] used space-time coupled spectral element method to solve the second-order hyperbolic equation. However, Chebyshev spectroscopy is rarely used to analyze vibration problems.

In this chapter, Chebyshev spectral element method is used to analyze the vibration problem under arbitrary load. By using the stable approximate vibration solution function of gravity center Lagrange interpolation and the flexibility of finite element method, two convergence modes of H convergence and P convergence are obtained for the vibration problem, and are compared with the matching point method.

2.1 The spectral element method

The spectral element method [42] was proposed by Patera in 1984 and applied to fluid dynamics. It combines the flexibility of the finite element method in dealing with boundary and structure with the quick convergence of the spectral method. The spectral element method can reduce the computational cost by using fewer units when the same precision is required. Within a cell, the spectral element method discretizes time into grid points corresponding to the zero points of the GLL polynomial, and performs Lagrange interpolation on these points.

The method for interpolation at a certain number of points means that in theory, when these points correspond to the zero points of the orthogonal polynomial, the interpolation accuracy is the highest. [86] As shown in Figure 2-1, uniform point distribution and GLL point distribution are close to the Runge function [see Equation (2.1)]. In the finite element method with uniform distribution, with the increase of interpolation times, the approximate numerical error becomes very large or the approximation fails. In this case, GLL interpolation has obvious advantages. The distribution of GLL points in the standard interval is shown in Figure 2-2.

$$f(x)=\frac{1}{1+25x^2} \tag{2.1}$$

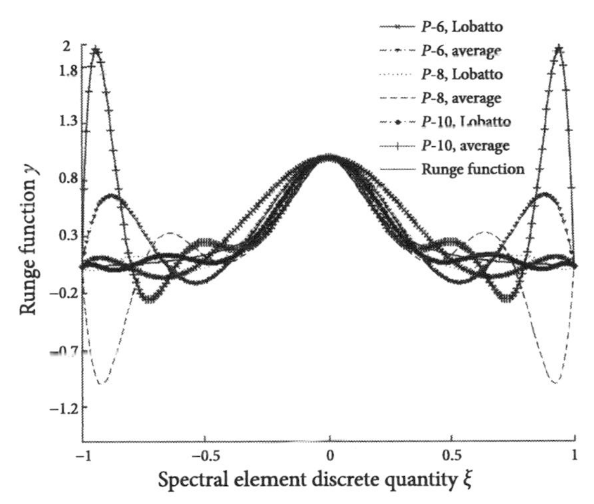

Figure 2-1 Approximate Runge function for interpolation of uniform points and GLL points

Differential equations, including dynamical equations, are commonly used to describe physical phenomena in nature. The difference method is discretized directly based on these differential equations or equations. But in most cases, the same physical process or phenomenon can be described in different forms. From the conservation law in the physical sense, the variational principle can be derived. And the variational problem is equivalent to the differential equation problem in some sense. The spectral element method is a discrete computing method based on the variational principle. The initial value problems of first-order linear differential equations are discussed below [See Equation (2.2)]. Higher-order linear differential equations can be converted into systems of first-order linear differential equations.

$$\begin{cases} \dfrac{\mathrm{d}t}{\mathrm{d}x} + A_s x = f(x,t) \\ x(0) = x_0 \end{cases} \tag{2.2}$$

In the formula, x is the time-dependent state variable, and $x \in \mathbf{R}^{N_v}$, N_v is the number of state variables; $f(x,t)$ is a function of the state variables x and time t; A_s is the coupling matrix of state variables, independent of time.

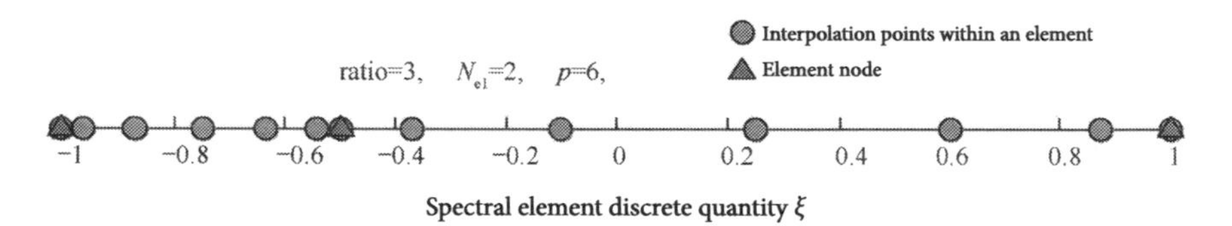

Figure 2-2 Distribution of GLL points in the standard interval

Note: Ratio is the ratio of the last unit to the first unit; N_{el} is the number of cells; p represents the number of interpolation nodes in a cell.

2.1.1 Transient and steady-state response analysis

The spectral element method is applied to discretize each state variable within a given time period and approximate to m-order Lagrange polynomial:

$$\tilde{x}^{(j)}(\xi) = \sum_{k=0}^{m} x^{(j)}(\xi_k) P_k^{(j)}(\xi) \tag{2.3}$$

In the formula, $p_k^{(j)}(\xi)$ is the k-th m-th Lagrange polynomial of the j-th element; ξ_k is the GLL point defined on [−1, 1]; $x^{(j)}(\xi_k)$ is the value of the unknown node at GLL point on the j-th cell. Lobatto polynomials are orthogonal polynomials defined by the differential of Legendre polynomials, see literature [86] page 146–151. Among them, $\xi \in [-1,1]$ is obtained through the mapping of formula (2.4).

$$x(\xi) = \frac{1}{2}\left[(x_2^{(j)} + x_1^{(j)}) + \frac{1}{2}(x_2^{(j)} - x_1^{(j)})\right]\xi \tag{2.4}$$

In the formula, $x_1^{(j)}$ and $x_2^{(j)}$ are the first node and the second node of the j-th unit respectively; ξ is the variable after domain transformation, and $\xi \in [-1,1]$. This field is shown in Figure 2-3.

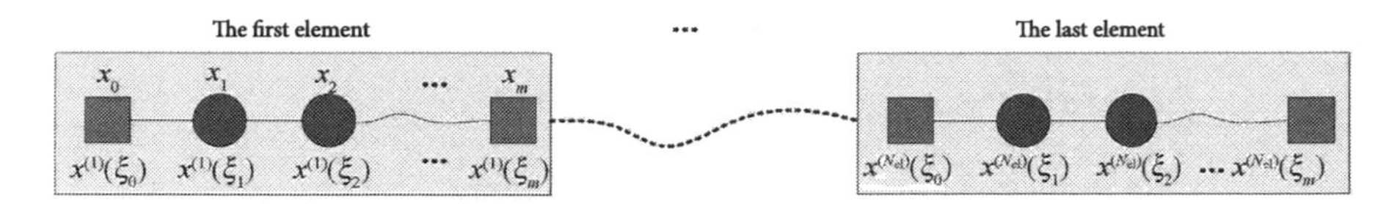

Figure 2-3 Discrete time domain is composed of spectral units, each of which is based on the M-order Lagrange polynomial of GLL point

Substituting equation (2.3) into equation (2.2), applying the Bubnov-Galerkin method [87] to each unit to minimize the interpolation error, the following equation can be obtained.

$$\sum_{j=1}^{N_{el}} \int_{-1}^{1} P_n^{(j)} \left\{ \frac{\mathrm{d}\tilde{x}^j}{\mathrm{d}\xi} + \frac{h^{(j)}}{2} \left[\boldsymbol{A}_s \tilde{x}^{(j)} - f^{(j)}\left(\tilde{x}^{(j)}, \xi\right) \right] \right\} \mathrm{d}\xi = 0 \tag{2.5}$$

In the above formula, $n = 0, 1, \cdots, m$; $h^{(j)}$ is the length of the j-th element; $P_n^{(j)}$ is the basic function of the n-th interpolation of Lagrange polynomials of order m. For each part of the equation (2.5) can obtain

$$\sum_{j=1}^{N_{el}} \left\{ \tilde{x}^{(j)} P_n^{(j)} \Big|_{-1}^{1} - \int_{-1}^{1} \left[\mathrm{d}\tilde{x}^{(j)} \frac{\mathrm{d}P_n^{(j)}}{\mathrm{d}\xi} + \frac{h^{(j)}}{2} \left(\boldsymbol{A}_s \tilde{x}^{(j)} P_n^{(j)} - f^{(j)}\left(\tilde{x}^{(j)}, \xi\right) P_n^{(j)} \right) \right] \mathrm{d}\xi \right\} = 0 \tag{2.6}$$

The integral in formula (2.6) is obtained by the Gauss-lobatto quadrature formula [see formula (2.7)].

$$\int_{-1}^{1} I \mathrm{d}\xi = \sum_{q=0}^{m} I(\xi_q)\omega_q \tag{2.7}$$

In the formula, I is the general function of ξ; ω_q is the weight [87] of Gauss-lobatto integral at q point. Each cell can be expressed as a matrix

$$\boldsymbol{\Phi} \begin{bmatrix} x(\xi_0) \\ x(\xi_1) \\ \vdots \\ x(\xi_m) \end{bmatrix}^{(j)} = \boldsymbol{A}_s \boldsymbol{I}_{\omega}^{(j)} \begin{bmatrix} x(\xi_0) \\ x(\xi_1) \\ \vdots \\ x(\xi_m) \end{bmatrix}^{(j)} - \boldsymbol{I}_{\omega}^{(j)} f^{(j)} \tag{2.8}$$

Among them,

$$\boldsymbol{\Phi}=\begin{bmatrix} \left.\frac{\mathrm{d}P_0}{\mathrm{d}\xi}\right|_{\xi_0}\omega_0+1 & \left.\frac{\mathrm{d}P_0}{\mathrm{d}\xi}\right|_{\xi_1}\omega_1 & \cdots & \left.\frac{\mathrm{d}P_0}{\mathrm{d}\xi}\right|_{\xi_m}\omega_m \\ \left.\frac{\mathrm{d}P_1}{\mathrm{d}\xi}\right|_{\xi_0}\omega_0 & \left.\frac{\mathrm{d}P_1}{\mathrm{d}\xi}\right|_{\xi_1}\omega_1 & \cdots & \left.\frac{\mathrm{d}P_1}{\mathrm{d}\xi}\right|_{\xi_m}\omega_m \\ \vdots & \vdots & \ddots & \vdots \\ \left.\frac{\mathrm{d}P_m}{\mathrm{d}\xi}\right|_{\xi_0}\omega_0 & \left.\frac{\mathrm{d}P_m}{\mathrm{d}\xi}\right|_{\xi_1}\omega_1 & \cdots & \left.\frac{\mathrm{d}P_m}{\mathrm{d}\xi}\right|_{\xi_m}\omega_m-1 \end{bmatrix} \tag{2.9}$$

$$\boldsymbol{I}_\omega=\frac{h^{(j)}}{2}\begin{bmatrix} \omega_0 & 0 & \cdots & 0 \\ 0 & \omega_1 & \cdots & 0 \\ \vdots & \vdots & \ddots & \vdots \\ 0 & 0 & \cdots & \omega_m \end{bmatrix} \tag{2.10}$$

Since two adjacent cells share one of the elements, it should be satisfied

$$x^{(j)}(\xi_m)=x^{(j+1)}(\xi_0) \tag{2.11}$$

Through Equation (2.8) and using the connection matrix [86] C, all spectral units of a state variable are assembled, the Galerkin approximate equation of a state variable is obtained:

$$\boldsymbol{B}_u\boldsymbol{X}_u=\boldsymbol{A}_\mathrm{s}\boldsymbol{B}_\omega\boldsymbol{X}_u-\boldsymbol{B}_\omega\boldsymbol{F}(\boldsymbol{X}_u) \tag{2.12}$$

In Equation (2.12), $\boldsymbol{B}_u$ and $\boldsymbol{B}_\omega$ are global differential matrix and global weight matrix; $\boldsymbol{F}(\boldsymbol{X}_u)$ is the global form of the incentive force, among them

$$\boldsymbol{X}_u=\begin{bmatrix} x|_{t_0} & x|_{t_1} & \cdots & x|_{t_{m\times N_\mathrm{el}+1}} \end{bmatrix}^\mathrm{T} \tag{2.13}$$

is all time node variables of state variable x expressed by spectral element method. The initial conditions of Equation (2.12) are processed, the first element of the first row and first column of $\boldsymbol{B}_u$ is set as 1, and the remaining elements of the first row and first column of $\boldsymbol{B}_u$ are set as 0. Then set the first element of $\boldsymbol{B}_\omega$ to 0, and the first element of $-\boldsymbol{B}_\omega\boldsymbol{F}(\boldsymbol{X}_u)$ the initial value in Equation (2.2), namely $x|_{t=0}=x_0$.

The spectral dispersion is the same as that of transient response analysis, but the global differential matrix and global weight matrix are different. Since the initial condition does not affect the steady-state response, it is not necessary to deal with the initial condition. Due to the particularity of the period, the first node of the first cell and the later node of the latter cell is required to be equal, that is

$$x^{(1)}(\xi_0) = x^{(N_{el})}(\xi_m) \tag{2.14}$$

The order for this equation is to first process the first row, the first column, the next row, and the last column. Then, to add the last row of the global differential matrix $\boldsymbol{B}_{\mathrm{u}}$ to the first row, add the last column to the first column, and then remove the last row and the next column. The global weight matrix $\boldsymbol{B}_{\mathrm{w}}$ does the same thing. For $F(\boldsymbol{X}_u)$, add the latter element to the first element, remove the latter element, and $\boldsymbol{X}_u$ becomes

$$\boldsymbol{X}_u = \left[x\big|_{t_0} \quad x\big|_{t_1} \quad \cdots \quad x\big|_{t_m \times N_{el}} \right]^{\mathrm{T}} \tag{2.15}$$

2.1.2 The global assembly and solution of all state variables

For N_v state variables, the global assembly formula of all state variables is obtained by the tensor cross multiplication of the coupling matrix A_s (the square matrix of $N_v \times N_v$):

$$(\boldsymbol{I} \otimes \boldsymbol{B}_u)\boldsymbol{X}_{ug} = (\boldsymbol{A}_s \otimes \boldsymbol{B}_\omega)\boldsymbol{X}_{ug} - (\boldsymbol{I} \otimes \boldsymbol{B}_\omega)\boldsymbol{F}_{ug}(\boldsymbol{X}_{ug}) \tag{2.16}$$

In the formula, $\boldsymbol{I}$ is $N_v \times N_v$ unit matrix; $\boldsymbol{X}_{ug}$ is a collection of all state variables at the time node. For transient response $\boldsymbol{X}_{ug}$, there are

$$\boldsymbol{T} = \begin{bmatrix} \begin{pmatrix} x_1\big|_{t_0} \\ x_1\big|_{t_1} \\ \vdots \\ x_1\big|_{t_m \times N_{el}+1} \end{pmatrix} \cdots \begin{pmatrix} x_{N_{el}}\big|_{t_0} \\ x_{N_{el}}\big|_{t_1} \\ \vdots \\ x_{N_{el}}\big|_{t_m \times N_{el}+1} \end{pmatrix} \end{bmatrix} \tag{2.17}$$

Simplified formula (2.16)

$$\boldsymbol{G}\boldsymbol{X}_{ug} = -\boldsymbol{B}_{\omega g}\boldsymbol{F}_{ug}(\boldsymbol{X}_{ug}) \tag{2.18}$$

$$\boldsymbol{G} = \boldsymbol{B}_{ug} - \boldsymbol{A}_{ug} \tag{2.19}$$

2.2 Chebyshev spectral element method for vibration analysis under arbitrary loads

The spectral approximation can choose the interpolation times freely and obtain p-convergence, while the finite element approximation can deal with the complex design domain flexibly and select the element size freely and obtain h-Convergence. The spectral element approximation combines the advantages of spectral approximation and finite element approximation.

2.2.1 The vibration problem and its integral form

The general form of considering vibration problem is

$$\dot{x} + A_{\mathrm{r}}x = f \tag{2.20}$$

Where A_{r} is the correlation matrix, which is related to mass, damping, and stiffness, and is assumed to be independent of time t; x and f are functions of time t.

In the Chebyshev spectral element method, to obtain the numerical solution of the vibration problem, a weight function W is introduced by using the Bubnov-Galerkin method, which is multiplied by both sides of Equation (2.20) at the same time, and integrated in the time domain. The integral form of vibration problem is obtained

$$\int_T Wx\mathrm{d}t + \int_T A_r Wx\mathrm{d}t = \int_T Wf\mathrm{d}t \tag{2.21}$$

In the formula, T represents the time domain.

2.2.2 Time unit division

As a kind of finite element method, the solution space Ω can be divided into N_e a non-overlapping unit space, namely

$$\boldsymbol{\Omega} = \bigcup_{e=1}^{N_e} \boldsymbol{\Omega}_e, \quad \bigcap_{e=1}^{N_e} \boldsymbol{\Omega}_e = \varnothing \tag{2.22}$$

The spectral element method through in each unit $\boldsymbol{\Omega}_e$ spectrum extension to approximate a function. The unit node basis function as shape functions in the unit $\boldsymbol{\Omega}_e$, vibration displacement can be approximate

$$\tilde{\boldsymbol{x}}^{(e)}(\xi)=\sum_{i=1}^{N_{\mathrm{esol}}}\boldsymbol{x}_i^{(e)}(\xi_i)\boldsymbol{\varphi}_i^e(\xi) \tag{2.23}$$

Type of vibration displacement of $\tilde{x}^{(e)}$ said unit Ω_e approximation functions; $x_i^{(e)}$ said unit Ω_e the ith node displacement values; Factor that had defined on the unit Ω focus though laser node basis function; $\varphi_i^e(\xi)$ is the base function of the gravity center Lagrange node defined on the element Ω_e; N_{esol} represents the number of solution nodes of each cell.

2.2.3 The discretization of vibration differential equation

Next, we use Chebyshev polynomials of the second kind to construct node basis functions. On the standard interval [−1, 1], the N-order node basis function can be expressed as Lagrange interpolation polynomial, which will pass through N+1 Chebyshev Gauss lobatto points, namely

$$\xi_j=-\cos\frac{j\pi}{N},\ j=1,2,\cdots,N+1 \tag{2.24}$$

Using the central interpolation formula, the Lagrange interpolation polynomial can be expressed as

$$b_i(\xi)=\frac{\dfrac{w_i}{\xi-\xi_i}}{\sum\limits_{j=1}^{N+1}\dfrac{w_j}{\xi-\xi_j}} \tag{2.25}$$

$$w_j=(-1)^{j-1}\delta_j,\quad \delta_j=\begin{cases}\dfrac{1}{2}, & j=1 \text{ or } j=N\\ 1, & \text{others}\end{cases} \tag{2.26}$$

It can be seen from Figure 2-4 that the characteristic of Kronecker δ of the node base function (if the two are equal, its output value is 1, otherwise it is 0), which ensures that the expansion coefficient $x_i^{(e)}$ in Equation (2.23) is consistent with the node value and that the boundary conditions are applied.

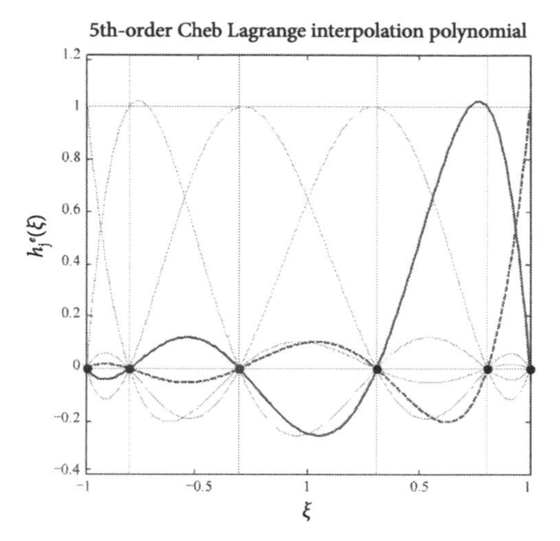

Figure 2-4 Interpolation basis function of Chebyshev Lagrange of order 5

To obtain the nodal basis function of the general element $\boldsymbol{\Omega}_e$, the nodal coordinate transformation is needed. The relationship between node basis function in standard element $\boldsymbol{\Omega}_{st}$ and general element $\boldsymbol{\Omega}_e$ can be expressed as

$$\nabla_x \boldsymbol{\varphi}_i^e\left[x(\xi)\right] = \boldsymbol{J}^{-1}\nabla_\xi \boldsymbol{h}_i^e(\xi) \tag{2.27}$$

In the formula, $x = x(\xi)$ defines the transformation of coordinates from general element $\boldsymbol{\Omega}_e$ to standard element $\boldsymbol{\Omega}_{st}$; ∇_x is a gradient operator about x; ∇_ξ is the gradient operator of ξ; $\boldsymbol{J}$ is the Jacobian matrix.

In this study, one-dimensional coordinate transformation can be expressed as

$$x(\xi) = \frac{1}{2}\left[(b-a)\xi + (b+a)\right] \tag{2.28}$$

In the formula, $\xi \in [-1,1]$, $x \in [a,b]$. The Jacobian matrix is a constant of $(b-a)/2$.

Substitute Equation (2.23) into Equation (2.21), the weight function is $\boldsymbol{W}_j(\xi)$, and then you integrate over the entire time domain, and you get

$$\begin{aligned}&\sum_{j=1}^{N_{\mathrm{sol}}} \boldsymbol{x}_j^{(e)}(\xi_i)\int_{-1}^{1}\left[\frac{\boldsymbol{\varphi}_j(\xi)}{\mathrm{d}\xi}\boldsymbol{W}_j(\xi) + \boldsymbol{A}_r\boldsymbol{J}_j\boldsymbol{\varphi}_j(\xi)\boldsymbol{W}_j(\xi)\right]\mathrm{d}\xi \\ &= \sum_{j=1}^{N_{\mathrm{sol}}} f_j^{(e)}(\xi_i)\int_{-1}^{1}\left[\boldsymbol{J}_j\boldsymbol{W}_j(\xi)\right]\mathrm{d}\xi;\ j = 1,2,\cdots,N_{\mathrm{sol}}\end{aligned} \tag{2.29}$$

When the weight function is the approximate node basis function of the spectral element, that is, $W_i(\xi) = \varphi_i(\xi)$, this spectral element method is called Galerkin spectral element method. When the weight function is the expression on the left side of the vibration differential equation, this spectral element method is called the least square spectral element method. The Galerkin element method was used in this study. Substitute φ_j as a weight function into equation (2.29) and convert it into a system of linear equations

$$\boldsymbol{KX} = \boldsymbol{F} \tag{2.30}$$

Among them,

$$\boldsymbol{K}_{ij} = \int_{-1}^{1}\left[\frac{\boldsymbol{\varphi}_j(\xi)}{\mathrm{d}\xi} + \boldsymbol{A}_s\boldsymbol{J}_j\boldsymbol{\varphi}_j(\xi)\right]\boldsymbol{\varphi}_i(\xi)\mathrm{d}\xi \tag{2.31}$$

$$\boldsymbol{F} = \sum_{j=1}^{N_{\mathrm{sol}}} \boldsymbol{f}_j^{(e)}(\xi_i)\int_{-1}^{1}\boldsymbol{J}_j\boldsymbol{\varphi}_j(\xi)\mathrm{d}\xi \tag{2.32}$$

$\boldsymbol{K}$ and $\boldsymbol{F}$ can be written as

$$\boldsymbol{K} = \boldsymbol{A} + \boldsymbol{A}_s\boldsymbol{JB} \tag{2.33}$$

$$\boldsymbol{F} = \boldsymbol{JDS} \tag{2.34}$$

Among them,

$$\boldsymbol{A} = \int_{-1}^{1}\frac{\boldsymbol{\varphi}_j(\xi)}{\mathrm{d}\xi}\boldsymbol{\varphi}_i(\xi)\mathrm{d}\xi \tag{2.35}$$

$$\boldsymbol{B} = \int_{-1}^{1}\boldsymbol{\varphi}_j(\xi)\boldsymbol{\varphi}_i(\xi)\mathrm{d}\xi \tag{2.36}$$

$$\boldsymbol{D} = \int_{-1}^{1}\boldsymbol{\Phi}_i(\xi)\mathrm{d}\xi \tag{2.37}$$

The matrix $\boldsymbol{A}$, $\boldsymbol{B}$, and $\boldsymbol{D}$ can be obtained by the Gauss-Chebyshev-Lobatto integral formula. [88]

2.2.4 Imposition of boundary conditions

The initial condition of velocity means that we must force the first row and first column of K to be equal to zero except for the first element, and force the first element of F to be equal to the initial value of velocity.

The initial condition of displacement means that the $(N+1)$ row and $(N+1)$ column of K force all elements except the $(N+1, N+1)$ to be equal to zero, and the $(N+1)$ element of F force to be equal to the initial displacement.

2.3 Application of clustering element spectral element method in dynamic analysis of structures under impact load

In engineering, nearly all mechanical structures are subjected to dynamic load, and in most cases, they will bear an impact load, such as with a hammer percussion screw, acoustic resonance percussion hammer percussion is detected in nondestructive testing components, automobile collisions, ship collisions with bridges, offshore platforms, plane crashes, surface vessels subjected to non-contact underwater explosion shock, [89] etc, to evaluate dynamic characteristics and behavior. This is why it is necessary to carry out simulation analysis. At present, there are many types of research on the dynamic behavior of the structure under impact load at home and abroad, such as the dynamic characteristic analysis of crane frame under lifting impact load, [90] the dynamic stress change study of gear under impact load, [91] and the dynamic response analysis of honeycomb sandwich plate under impact load. [92] Many papers have analyzed and studied the dynamic characteristics and dynamic behavior of various structures under impact load, but there are no relevant reports on the dynamic analysis of structures under impact load by spectral element method.

The spectral element method is a numerical method applied to CFD proposed by Patera in 1984, which has the advantages of flexibility of finite element method in dealing with arbitrary structures and their boundaries and rapid convergence of spectral methods [81]. The spectral element method can obtain the same precision as other methods with fewer elements. Its characteristic is to discretization each element at the zero points of GLL and then Lagrange polynomial interpolation. Theoretically, the highest interpolation accuracy can be obtained when interpolation is at the zero-point of orthogonal polynomial. [86] The author from the perspective of size calculation has presented this book based on the spectral element method of structure dynamic response simulation time through a step by step approach. This could be divided into the short time period, the simulation time in each time period partition unit, where each time period will be considered a separate part, and the results will be part of the former as part of the calculation of the initial conditions. This method saves calculation time [53]. At the zero-point of the second Chebyshev orthogonal polynomial, a discrete scheme for nonlinear vibration problem [54] is constructed from the Angle of the Lagrange interpolation of gravity center. The key time points in the dynamic response optimization are calculated by taking advantage of the high precision of the discrete interpolation of spectral elements. [38] M. H. Kurdi [93] solved a simple mass-spring damping system by using the temporal spectral element method, and optimized the single-degree-of-freedom vibration absorber and single-degree-of-freedom micro-controller on this basis.

In this chapter, based on the governing equation of structural dynamics, the finite element method is used for spatial discretization and the spectral element method for time discretization. Given the short impact load time and large variation, the discrete spectral elements are aggregated to make up for the defect of large isometry errors.

The spectral element method is divided into temporal spectral element method, space spectral element method and time-space coupled spectral element method. For dynamic equations, you can use either of these methods. The following is an example of the time spectrum method.

The main steps of the time spectrum method are: a) The dynamic equation is transformed into a set of first-order linear differential equations, and then the Bubnov-Galerkin method is equivalent to the integral form; b) According to the characteristics of the impact load, the simulation time is divided into aggregation units, that is, the size of the unit in the position of sudden load is relatively small, and the size of the unit in the position of flat load is relatively large, which can be controlled by the ratio of the maximum size and the minimum size of the unit; c) Each time unit is divided into several time nodes, namely the zero-point of orthogonal polynomial; d) The approximate solution of each element is represented by a linear combination of orthogonal polynomials by Legendre; e) The connection matrix is used to integrate the approximate solutions of all elements into the general spectral element equation; f) The global displacement approximate solution and the velocity approximate solution can be obtained by solving the general spectral element equation. The acceleration solution can be obtained by solving the dynamic differential equation.

2.3.1 Dynamic response equation of a linear structure and its transformation form

The dynamic response equation of linear structure can be expressed as

$$\boldsymbol{M}\ddot{\boldsymbol{x}} + \boldsymbol{C}\dot{\boldsymbol{x}} + \boldsymbol{K}\boldsymbol{x} = \boldsymbol{F} \tag{2.38}$$

In the formula, $\boldsymbol{M}$ is the mass matrix; $\boldsymbol{C}$ is the damping matrix; $\boldsymbol{K}$ is the stiffness matrix; $\boldsymbol{F}$ is the dynamic load vector; $\boldsymbol{x}$ is the displacement vector; $\ddot{\boldsymbol{x}}$ is the acceleration vector, $\dot{x}$ is the velocity vector. The initial conditions are $x(0) = b_0$, $\dot{x}(0) = v_0$. In Equation (2.38), M, C and K do not change with time; x, $\ddot{x}$, $\dot{x}$ are functions of time; F is an arbitrary time function, time $t \in [t_0, t_n]$.

To the spectral element method is used for, $x_1 = \dot{x}$, $x_2 = x$, type (2.38) can be converted into a first-order linear ordinary differential equation [see type (2.39)], the solution of the equation with type (2.38).

$$\begin{cases} \begin{bmatrix} \dot{x}_1 \\ \dot{x}_2 \end{bmatrix} + \begin{bmatrix} M & 0 \\ 0 & I \end{bmatrix}^{-1} \begin{bmatrix} C & K \\ -I & 0 \end{bmatrix} \begin{bmatrix} x_1 \\ x_2 \end{bmatrix} = \begin{bmatrix} M & 0 \\ 0 & I \end{bmatrix}^{-1} \begin{bmatrix} F \\ 0 \end{bmatrix} \\ x_1(0) = x_1^0 x_2(0) = x_2^0 \end{cases} \tag{2.39}$$

2.3.2 Aggregation unit division

As shown in Figure 2-5, simulation time $t \in [t_0, t_n]$ is divided into N disjoint cells, that is $[t_0, t_1]$, $[t_1, t_2]$, $[t_2, t_3]$, ···, $[t_{(n-1)}, t_n]$, each cell is configured with several points. In the division of units, the maximum point of impact load is taken as the center, and the unit size at the center is the smallest. The larger the unit size is, the larger the unit size is. Each cell can be configured with Chebyshev or Legendre orthogonal polynomials for the same or the different number of points, and the same zero configuration is used for this chapter. In this way, the local mutation of impact load can be avoided. In the specific implementation, ratio $= \dfrac{l_{\frac{n}{2}+1}}{l_n} = \dfrac{l_{\frac{n}{2}-1}}{l_1}$ (ratio = 0.1 in Figure 2-5). Figure 2-5 (a) shows the aggregation unit schematic diagram, and Figure 2-5 (b) shows the corresponding impact load of 30 aggregation units, in which the impact load is controlled by three parameters, namely the maximum value, width, and load center of the impact load.

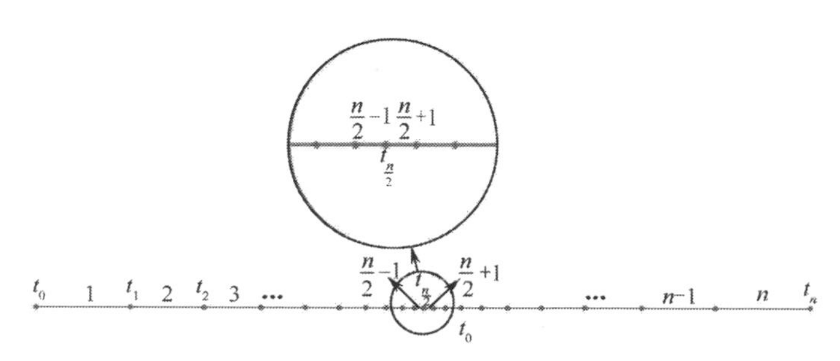

(a) Schematic diagram of aggregation unit

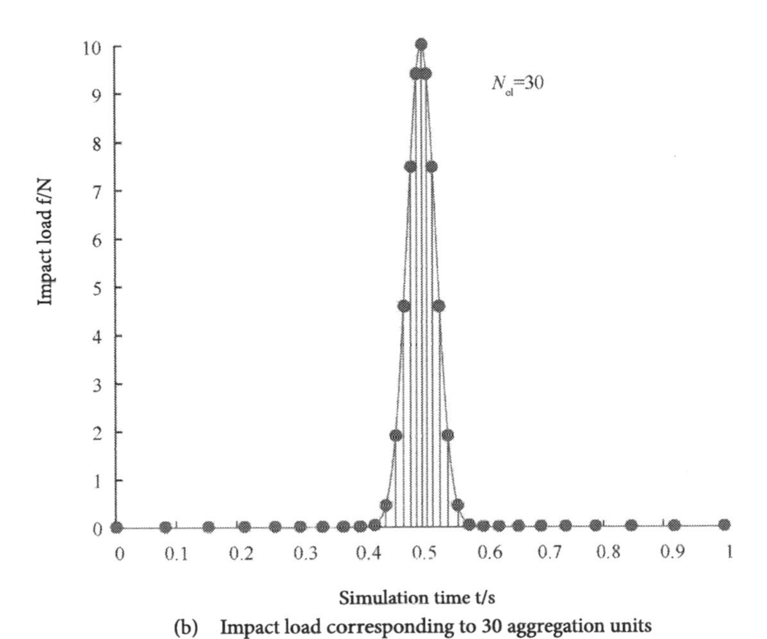

(b) Impact load corresponding to 30 aggregation units

Figure 2-5 Aggregation unit and corresponding impact load under temporal spectral element method

2.3.3 Element analysis

After the aggregate element is divided, the zeros of orthogonal polynomials are added in the element interval, and the approximation accuracy can be improved by adding interpolation points, which is called HP method within the finite element method.

In the temporal spectral element method, the interpolation GLL points are arranged at special positions in the element, and the basis function can be expressed as the combination of orthogonal polynomials, so as to form the shape function of each point in the element. In this way, interpolation can be carried out on a limited number of points to achieve the convergence speed of the spectral method.

Two orthogonal polynomials, namely Chebyshev and Legendre, are generally used for the time element method, and Legendre orthogonal polynomials are used for this chapter [86].

Since the cell endpoints are not zero points of the orthogonal polynomials, a Lobbato polynomial is added to ensure that the cell endpoints are included at zero points. Lobbato polynomial satisfies orthogonal characteristics:

$$\int_{-1}^{1}(1-\xi^2)L_{oi}(\xi)L_{oj}(\xi)\mathrm{d}\xi = \frac{2(i+1)(i+2)}{2i+3}\delta_{ij} \tag{2.40}$$

In the formula, δ_{ij} is Kronecker δ function;

$$L_{oj}(\xi)=\sum_{k=0}^{m}x^{(j)}(\xi_k)P_k^{(j)}(\xi)$$

$$L_{oi}(\xi)=\sum_{k=0}^{m}x^{(i)}(\xi_k)P_k^{(i)}(\xi)$$

The GLL point and its weight can be obtained by the solution (2.40).

$$\left(1-\xi^2\right)P_N^{'}(\xi)=0 \tag{2.41}$$

Where N is the times of interpolation, and the weight can be expressed as

$$\omega_k=\frac{2}{N(N+1)}\frac{1}{L_N^2(\xi_k)} \tag{2.42}$$

In the type, zero point ξ_k to $L_{o_{N-1}}(\xi)$, $k = 1, 2, \cdots N-1$.

In the simulation time, the scheme is divided as shown in Figure 2.5, and the state variable is expressed as m-degree Lagrange polynomial in combination with the discrete impact load:

$$\tilde{x}^{(j)}(\xi)=\sum_{k=0}^{m}x^{(j)}(\xi_k)P_k^{(j)}(\xi) \tag{2.43}$$

In the formula, $P_k^{(j)}(\xi)$ is the polynomial of Lagrange degree k of unit j; ξ_k is the GLL point defined on $[-1, 1]$; $x^{(j)}(\xi_k)$ is the value of the unknown node on cell J at GLL. And then by the Bubnov-Galerkin method

$$\sum_{j=1}^{N_{el}}\int_{-1}^{1}P_{-1}^{(j)}\left[\frac{\mathrm{d}\tilde{x}^{(j)}}{\mathrm{d}\xi}+\frac{h^{(j)}}{2}\left\{A\tilde{x}^{(j)}-f^{(j)}(\tilde{x}^{(j)},\xi)\right\}\right]\mathrm{d}\xi=0 \tag{2.44}$$

The discretization of each element is expressed in the form of a matrix:

$$\boldsymbol{L}^e\boldsymbol{X}^e(t)=\boldsymbol{F}^e(t) \tag{2.45}$$

In the type, $\boldsymbol{L}^e=\boldsymbol{\Phi}-A_s\boldsymbol{I}_\omega^{(e)}$; $\boldsymbol{x}^e=\left(\left[x(\xi_0),x(\xi_1),\cdots,x(\xi_m)\right]^{(e)}\right)^{\mathrm{T}}$; $\boldsymbol{F}^e(t)=-\boldsymbol{I}_\omega^{(e)}f^{(e)}$, I_ω is a general function of ξ,

$$\boldsymbol{I}_\omega=\frac{h^{(j)}}{2}\begin{bmatrix}\omega_0 & 0 & \cdots & 0\\ 0 & \omega_1 & \cdots & 0\\ \vdots & \vdots & \ddots & \vdots\\ 0 & 0 & \cdots & \omega_m\end{bmatrix}$$

$$\boldsymbol{\Phi}=\begin{bmatrix} \left.\frac{\mathrm{d}P_0}{\mathrm{d}\xi}\right|_{\xi_0}\omega_0+1 & \left.\frac{\mathrm{d}P_0}{\mathrm{d}\xi}\right|_{\xi_1}\omega_1 & \cdots & \left.\frac{\mathrm{d}P_0}{\mathrm{d}\xi}\right|_{\xi_m}\omega_m \\ \left.\frac{\mathrm{d}P_1}{\mathrm{d}\xi}\right|_{\xi_0}\omega_0 & \left.\frac{\mathrm{d}P_1}{\mathrm{d}\xi}\right|_{\xi_0}\omega_1 & \cdots & \left.\frac{\mathrm{d}P_1}{\mathrm{d}\xi}\right|_{\xi_m}\omega_m \\ \vdots & \vdots & \ddots & \vdots \\ \left.\frac{\mathrm{d}P_m}{\mathrm{d}\xi}\right|_{\xi_0}\omega_0 & \left.\frac{\mathrm{d}P_m}{\mathrm{d}\xi}\right|_{\xi_1}\omega_1 & \cdots & \left.\frac{\mathrm{d}P_m}{\mathrm{d}\xi}\right|_{\xi_m}\omega_m-1 \end{bmatrix}$$

2.3.4 Integrated spectral element equation and solution

Integration means the combination of the spectral element equations of all elements in discrete order to obtain the overall spectral element equation. For a system containing N_v equations of state, the global assembly of all state variables can be obtained by the tensor cross product of the coupling matrix A_s (square matrix of $N_v \times N_v$):

$$(\boldsymbol{I}\otimes\boldsymbol{B}_u)\boldsymbol{X}_{ug}=(\boldsymbol{A}_s\otimes\boldsymbol{B}_\omega)\boldsymbol{X}_{ug}-(\boldsymbol{I}\otimes\boldsymbol{B}_\omega)\boldsymbol{F}_{ug}(\boldsymbol{X}_{ug}) \tag{2.46}$$

In the formula, $\boldsymbol{B}_u$ is the global differential matrix; $\boldsymbol{B}_\omega$ is the global weight matrix; $\boldsymbol{F}_{ug}(\boldsymbol{X}_{ug})$ is the global form of the incentive force; $\boldsymbol{X}_{ug}$ is the collection of all state variables at the time node. The simplification (2.46) gives

$$\boldsymbol{G}\boldsymbol{X}_{ug}=-\boldsymbol{B}_{\omega g}\boldsymbol{F}_{ug}(\boldsymbol{X}_{ug}) \tag{2.47}$$

In the formula, $\boldsymbol{G}$ is the global linear matrix of the time period; Equation (2.47) is a system of linear equations, which can be solved by the Gaussian elimination method.

2.4 Chebyshev spectral element method for nonlinear vibration analysis

Although many engineering problems can be approximated by linear vibration, many engineering vibrations need to be considered nonlinear. For example, large Angle single pendulum, vibration conveyor, vibration system formed by gas resistance of high-speed train and elastic-plastic deformation of materials, etc., [94] all need to be analyzed by a non-linear differential equation. Nonlinear vibration does not conform to the superposition principle and is usually analyzed numerically.

Orszag S. A. [95] proposed the spectral method in 1969, which brings hope to the high-precision numerical analysis concerned by researchers. However, its shortcomings, such as its inability to deal with complex design fields and its inability to approximate non-smooth functions, limit its development. Given the high precision and exponential convergence of the spectral method and the flexibility of the finite element method in dealing with boundary, scholar Patera proposed the spectral element method in 1984. This method constructed node basis function by Lagrange interpolation at GLL points and was applied to fluid dynamics analysis. [81] For more than 30 years, the spectral element method has attracted great attention due to its high precision and rapid convergence, and has been successfully applied in many fields of science and engineering. [96–101] In dynamic response optimization, the spectral element method satisfies dynamic constraint conditions by precisely solving the dynamic governing equation and combining GLL points to obtain better solutions. In mechanical fault diagnosis, the spectral element method can be used to simulate the guided wave excitation and reception of a three-dimensional plate structure with cracks, as well as wave propagation [102]. The simulation time is divided into several steps, and the stepwise time spectrum element method [53] is adopted to simulate the THREE-DIMENSIONAL cantilever beam, which can obtain the same results as ANSYS simulation, but the efficiency is higher than ANSYS simulation. Literature [38] applies the spectral element discrete scheme to the identification of critical time points of structural dynamic stress. Zhao et al. [103] used Chebyshev least-square spectral element method to analyze and solve the radiant heat transfer of translucent media in detail. Lin Weijun [104] used the Modal Basis spectral element method to elaborate the theoretical formula of elastic wave propagation simulation and applied Chebyshev orthogonal polynomial expansion. Geng Yanhui et al. [105] proposed the time-space coupled spectral element method and applied it to the solution of inhomogeneous one-dimensional, two-dimensional, and three-dimensional wave equations of the first class of boundary conditions. In reference, [106] the time-space Grllerkin spectral element method was applied to solve the Burgers equation with small viscosity. A sub-cycle technique in which the hyperbolic control equation was divided into the explicit method and the implicit method of a parabolic control equation was studied. In literature, [107] spectral element formulas of mixed time and space fields are proposed, explicit and implicit algorithms are developed, and used to solve second-order scalar hyperbolic equations. In literature, [108] the space-temporal spectral element method is used to solve the vibration problem of a simply supported modified Euler-Bernoulli nonlinear beam under forced lateral vibration.

In this chapter, the node basis function is constructed by Lagrange interpolation in the center of gravity of Chebyshev orthogonal polynomials, and the Chebyshev spectral element method for solving nonlinear vibration problems is proposed.

For the nonlinear term in the nonlinear vibration problem, it is firstly directly differentiated and then added into the discrete formula of the linear vibration problem, which is converted into Newton-Raphson iterative formula for the iterative solution.

The nonlinear equations can be expressed as

$$\frac{\mathrm{d}\boldsymbol{X}(t)}{\mathrm{d}t}=\boldsymbol{F}(\boldsymbol{X}(t)) \tag{2.48}$$

In the formula, $\boldsymbol{X}(t)$ is an n-dimensional solution vector, and $\boldsymbol{F}(\boldsymbol{X}(t))$ is an m-dimensional function vector.

Considering function F: $R^n \to R^m$, among them,

$$\boldsymbol{F}(x_1,x_2,\cdots,x_n)=\begin{bmatrix} f_1(x_1,x_2,\cdots,x_n) \\ f_2(x_1,x_2,\cdots,x_n) \\ \vdots \\ f_m(x_1,x_2,\cdots,x_n) \end{bmatrix}$$

So the Jacobian matrix of $F(x_1, x_2, \cdots, x_n)$ is

$$\boldsymbol{J}_F(x_1,x_2,\cdots,x_n)=\begin{bmatrix} \frac{\partial f_1}{\partial x_1} & \cdots & \frac{\partial f_1}{\partial x_n} \\ \vdots & \ddots & \vdots \\ \frac{\partial f_m}{\partial x_1} & \cdots & \frac{\partial f_m}{\partial x_n} \end{bmatrix}$$

Newton-Raphson iterative formula is expressed as

$$\boldsymbol{J}_F\Delta\boldsymbol{X}=\boldsymbol{F}(\boldsymbol{X}(t)) \tag{2.49}$$

In the type, $\Delta\boldsymbol{X}=\boldsymbol{X}_{i=1}-\boldsymbol{X}_i$.

2.5 Calculation example analysis

2.5.1 Analysis of vibration under arbitrary loads

1. Linear loading

The ordinary differential equation of linear load is

$$\ddot{x}+\dot{x}-2x=2t \tag{2.50}$$

The initial condition: $x(0)=0$, $\dot{x}(0)=0$. The exact solution is $x=e'-e^{-2t}/(2-t-1)$.

Figure 2-6 shows Chebyshev spectral element method solving the linear load vibration problem. On the interval [0, 6], for different elements and interpolation times, different schemes are adopted to calculate the maximum absolute displacement error of the linear load vibration problem, as shown in Table 2.1.

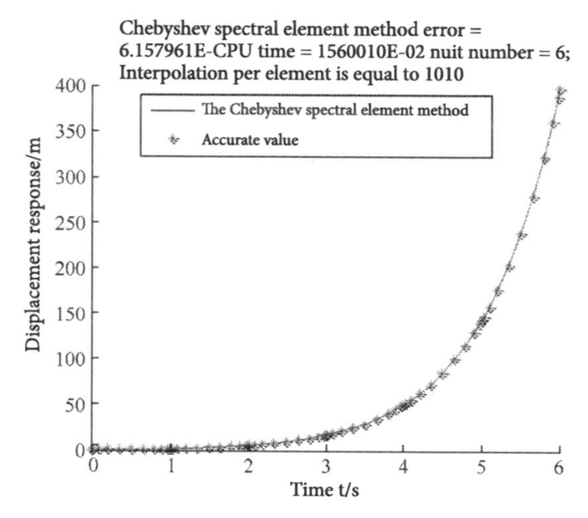

Figure 2-6 Chebyshev spectral element method is used to solve the vibration problem under linear loads

Table 2.1 The maximum absolute displacement error of linear load vibration problem is calculated by different schemes

Element number	Interpolation times per second	Maximum absolute displacement error (Chebyshev element method)	Element number	Interpolation times per second	Maximum absolute displacement error (collocation method)
10	10	4.939×10^{-11}	10	10	0.111515
10	20	3.086×10^{-11}	10	20	3.836×10^{-8}
20	50	5.996×10^{-12}	20	50	8.235×10^{-8}
10	100	1.207×10^{-10}	10	100	1.232×10^{-5}
5	300	3.816×10^{-10}	5	300	0.0030

(Continued)

Element number	Interpolation times per second	Maximum absolute displacement error (Chebyshev element method)	Element number	Interpolation times per second	Maximum absolute displacement error (collocation method)
5	500	7.202×10^{-10}	5	1000	0.566
50	100	5.900×10^{-10}	50	2000	3.283
100	50	2.637×10^{-11}	100	5000	396.928

It can be seen from Table 2.1 that the linear load vibration problem calculated by the Chebyshev spectral element method has high accuracy. The number of elements ranges from 10 to 100, the number of interpolations ranges from 10 to 500, and the maximum absolute error of displacement reaches 10–10 orders of magnitude. It can be further seen from Figure 2-7 that the method in this chapter has obtained high accuracy from both *h* convergence and *p* convergence, but *p* convergence is more stable. In the collocation method, the number of interpolations ranges from 10 to 5000, and the highest accuracy reaches the order of 10–8. However, as the number of interpolations increases, the error becomes larger and larger, and even wrong solutions are obtained. When the number of interpolations is 5000, the maximum absolute error of the displacement is 396.928.

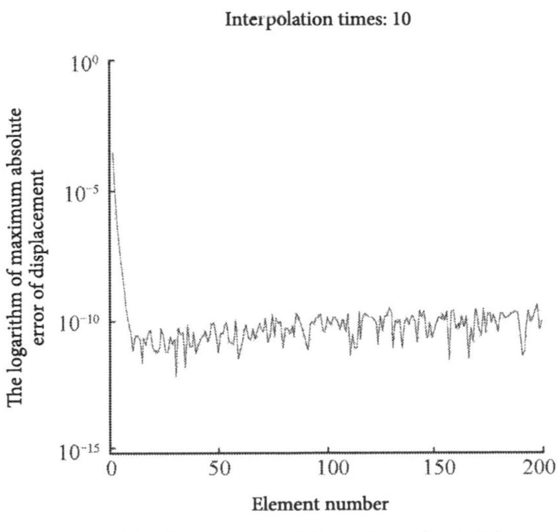

(a) *h* convergence (Element number is 10)

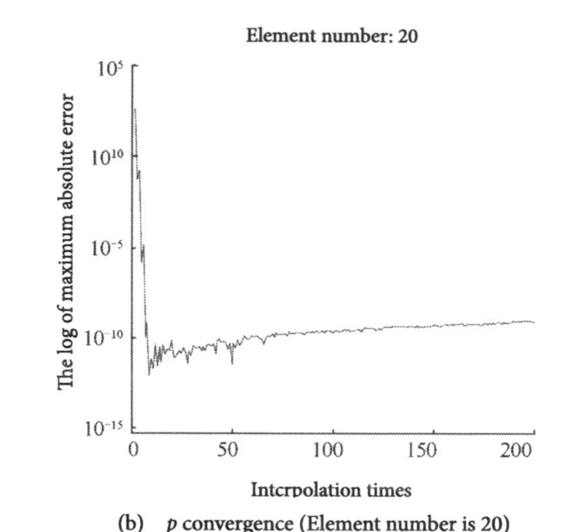

(b) p convergence (Element number is 20)

Figure 2-7 Two convergence curves of Chebyshev spectral element method for vibration problem under linear loading

2. *Triangular load*

The vibration differential equation of a linear undamped system can be described as follows

$$0.5\ddot{\boldsymbol{x}}+8\pi^2\boldsymbol{x}=\boldsymbol{F}(t) \tag{2.51}$$

The initial conditions are $x(0) = 0, \dot{x}(0) = 0$, and the trigonometric external force is

$$F(t)=\begin{cases}\dfrac{2\boldsymbol{F}_0}{t_0}t,\ 0<t\leqslant\dfrac{1}{2}t_0\\ -\dfrac{2\boldsymbol{F}_0}{t_0}(t-t_0),\ \dfrac{1}{2}t_0<t\leqslant t_0\\ 0,\ t>t_0\end{cases} \tag{2.52}$$

In the formula, $F_0 = 100, t_0 = 0.4$.

The exact solution of the vibration differential equation is

$$x(t)=\begin{cases}\dfrac{2\boldsymbol{F}_0}{k}\left[\dfrac{t}{t_0}-\dfrac{t_\omega}{2\pi t_0}\sin\left(2\pi\dfrac{t}{t_\omega}\right)\right],\ 0<t\leqslant\dfrac{1}{2}t_0\\ \dfrac{2\boldsymbol{F}_0}{k}\left\{1-\dfrac{t}{t_0}+\dfrac{t_\omega}{2\pi t_0}\left(2\sin\left[\dfrac{2\pi}{t_\omega}\left(t-\dfrac{1}{2}t_0\right)\right]-\sin\left(2\pi\dfrac{t}{t_\omega}\right)\right)\right\},\ \dfrac{1}{2}t_0<t\leqslant t_0\\ \dfrac{2\boldsymbol{F}_0}{k}\left\{\dfrac{t_\omega}{2\pi t_0}\left(2\sin\left[\dfrac{2\pi}{t_\omega}\left(t-\dfrac{1}{2}t_0\right)\right]-\sin\left(2\pi\dfrac{t}{t_\omega}\right)\right)\right\},\ t>t_0\end{cases} \tag{2.53}$$

In the formula, k and t_w respectively, represent the stiffness and vibration period of the system.

Different numbers of elements and interpolation times are used to analyze the displacement within 0–1.2 s, and the maximum absolute error of displacement for calculating the triangular load vibration problem is obtained by different schemes, as shown in Table 2.2. Figure 2-8 shows the use of the Chebyshev spectral element method to solve the triangular load vibration problem.

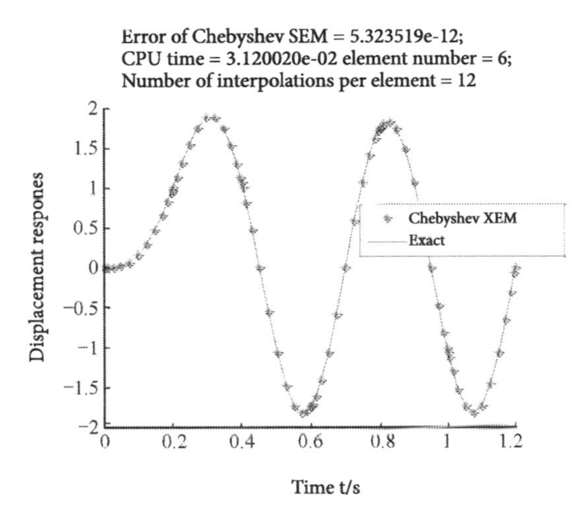

Figure 2-8 Chebyshev spectral element method was used to solve the triangular load vibration problem

Table 2.2 Calculation of maximum absolute displacement error of triangular load vibration under different schemes

Element number	Interpolation times per second	Maximum absolute displacement error (Chebyshev element method)	Element number	Interpolation times per second	Maximum absolute displacement error (collocation method)
10	10	0.0036	10	20	0.03118
6	100	9.925×10^{-14}	6	50	0.00778
12	300	4.146×10^{-12}	12	100	0.00015
18	12	7.549×10^{-15}	18	200	0.00050
18	50	2.711×10^{-13}	18	500	6.987×10^{-5}

(Continued)

Element number	Interpolation times per second	Maximum absolute displacement error (Chebyshev element method)	Element number	Interpolation times per second	Maximum absolute displacement error (collocation method)
24	11	3.019×10^{-14}	24	1000	5.346×10^{-6}
48	16	8.548×10^{-14}	48	2000	5.022×10^{-6}
48	50	7.007×10^{-13}	48	3000	4.267×10^{-6}

It can be seen from Table 2.2 that using the Chebyshev spectral element method to solve the vibration problem under the triangular load can obtain higher accuracy than solving the vibration problem under the linear load. When the number of units is 12 and the number of interpolations is 300, the maximum absolute error of displacement is 4.146×10^{-12}. At the same time, the smallest maximum absolute error is 7.549×10^{-15} when the number of units is 18 and the number of interpolations is 12. The largest absolute error is the unit When the number is 10 and the interpolation number is 10, it is 0.0036. The minimum displacement maximum absolute error obtained by the collocation method is 4.267×10^{-6}, and the maximum displacement maximum absolute error is 0.03118. Therefore, the Chebyshev spectral element method has higher accuracy than the collocation method. From Figure 2-9, it can be further illustrated that the Chebyshev spectral element method is not only accurate but also stable. Comparing Figure 2-9 (a) and Figure 2-9 (b), we can see that h convergence is more stable.

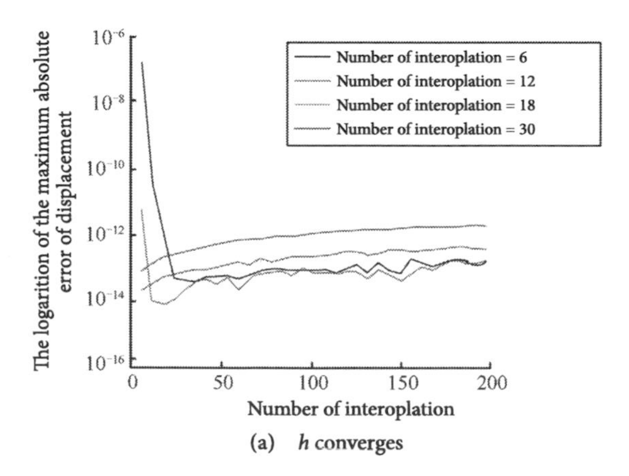

(a) *h* converges

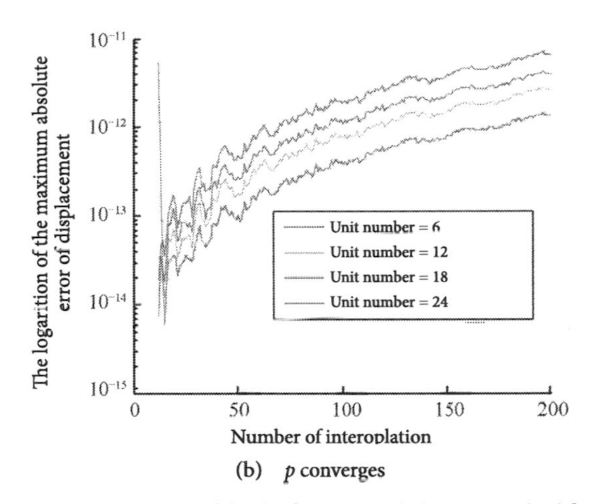

(b) p converges

Figure 2-9 Two convergence curves of Chebyshev spectral element method for vibration problem under triangular load

3. *Semi-sinusoidal pulse load*

The governing equation is

$$\dot{x} + \omega^2 x = f(t) \tag{2.54}$$

The initial conditions are $x(0) = 0, \dot{x}(0) = 0, \omega = 10\pi$. The pulse load of semi-sine wave is

$$f(t) = \begin{cases} \sin\dfrac{\pi t}{t_1}, t < t_1 \\ 0, \ t > t_1 \end{cases}$$

The exact solution is

$$x(t)\begin{cases} \dfrac{1}{\dfrac{\tau}{2t_1} - \dfrac{2t_1}{\tau}}\left(\sin\dfrac{2\pi t}{\tau} - \dfrac{2t_1}{\tau}\sin\dfrac{\pi t}{t_1}\right), t < t_1 \\ \dfrac{1}{\dfrac{\tau}{2t_1} - \dfrac{2t_1}{\tau}}\left[\sin\dfrac{2\pi t}{\tau} + \dfrac{2t_1}{\tau}\sin\left(\dfrac{t}{\tau} - \dfrac{t_1}{\tau}\right)\right], t > t_1 \end{cases} \tag{2.55}$$

When $t_1 = 10$, Chebyshev spectral element method is used to calculate the vibration problem under the impulse load of the semi-sinusoidal wave. Figure 2-10 and Table 2.3 show the obtained results.

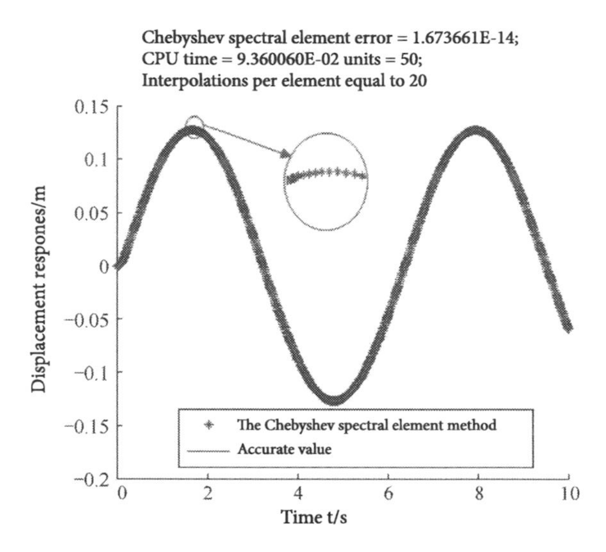

Figure 2-10 Chebyshev spectral element method is used to solve the vibration problem under semi-sinusoidal pulse load

Table 2.3 The maximum absolute displacement error of the vibration problem under semi-sinusoidal pulse load is calculated by different schemes

Element number	Interpolation times per second	Maximum absolute displacement error (Chebyshev element method)	Element number	Interpolation times per second	Maximum absolute displacement error (collocation method)
6	100	0.00018	6	40	0.00771
12	50	0.00017	12	50	0.00499
12	20	3.717×10^{-6}	12	90	0.00074
48	50	4.143×10^{-8}	48	160	0.00049
50	12	1.692×10^{-12}	50	200	0.00037
50	20	1.673×10^{-14}	50	240	0.00013
100	12	4.961×10^{-15}	100	600	1.823×10^{-6}

(Continued)

Element number	Interpolation times per second	Maximum absolute displacement error (Chebyshev element method)	Element number	Interpolation times per second	Maximum absolute displacement error (collocation method)
200	12	6.841×10^{-15}	200	2400	7.957×10^{-7}
300	12	9.374×10^{-15}	300	3200	1.401×10^{-5}
300	20	9.503×10^{-14}	300	6000	3.194×10^{-5}

It can be seen from Table 2.3 that for the Chebyshev spectral element method when the number of units is relatively small and the number of interpolations is relatively large, the accuracy of the results obtained is low. For example, when the number of units is 6 and the number of interpolations is 100, the maximum absolute error of displacement is 0.00018. If the number is relatively large and the number of interpolations is relatively small, the results obtained have high accuracy. For example, when the number of units is 100 and the number of interpolations is 12, the maximum absolute error of the displacement is 4.961×10^{-15}. For the collocation method, when the times of interpolation is 2400, the highest accuracy can be obtained, and the maximum absolute error of the displacement is 7.957×10^{-7}. When the times of interpolation is 40, the accuracy is the lowest, and the maximum absolute error of the displacement is 0.00771. In short, with the Chebyshev spectral element method, as long as the number of units and the times of interpolations are reasonably selected, an accuracy of more than 10^{-10} orders of magnitude can be obtained.

4. ***The cantilever beam***

In space, the traditional finite element method discretizes the differential equation and forms the quadratic differential equation on node deformation [7, 8]:

$$\boldsymbol{M}\{\ddot{w}\}+\boldsymbol{K}\{w\}=f(t) \tag{2.56}$$

In the formula, $\{w\}$ is node deformation. Each node includes 6 deformations, 3 displacements $\{x, y, z\}$ and 3 rotations $\{\theta_x, \theta_y, \theta_z\}$, where damping is not considered. The geometrical dimensions and physical parameters of the cantilever beam are shown in Table 2.4.

Table 2.4 Geometrical dimensions and physical parameters of the cantilever beam

Parameter	Number of values
Moment of inertia of section I_Z/cm^4	1450
Moment of inertia of section I_Y/cm^4	1269
Modulus of elasticity $E/(\text{N/cm}^2)$	10^7
Shear modulus $G/(\text{N/cm}^2)$	3.9×10^6
Beam cross-sectional area A/cm^2	19.5
The length of the beam L/cm	100
The density of the beam $\rho/(\text{kg/cm}^3)$	2.588×10^{-4}
The amplitude of the excitation force b/N	3.0×10^4

Euler beam element can be used to discrete the cantilever beam in space, and each node has 6 degrees of freedom. Here, the cantilever beam is discretized into 10 units, that is, 11 nodes, with the left end fixed, with a total of 60 degrees of freedom. Cantilever beam end force $f(t) = b\sin(\omega t)$, vertical upward direction.

When the load frequency ω was 10 π, 176.64 π, and 184 π, the vertical displacement response of the free end of the cantilever beam was solved by Chebyshev element method and compared with the ANSYS analysis, as shown in Figure 2-11.

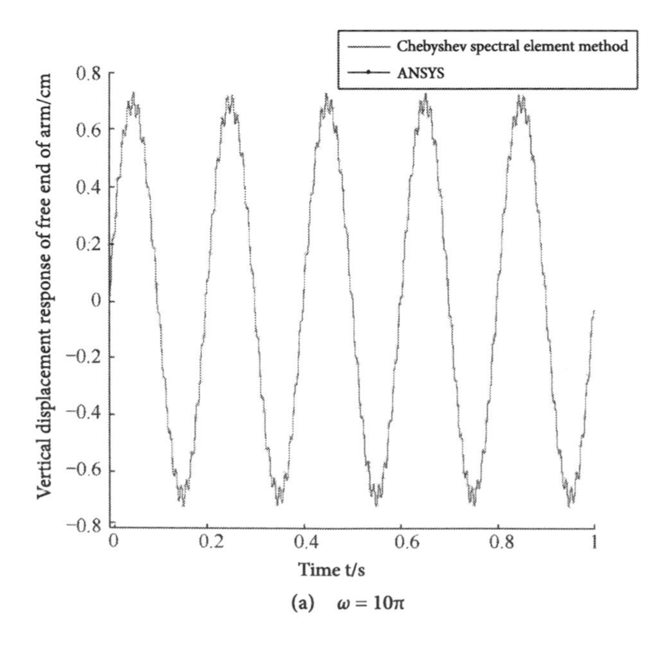

(a) $\omega = 10\pi$

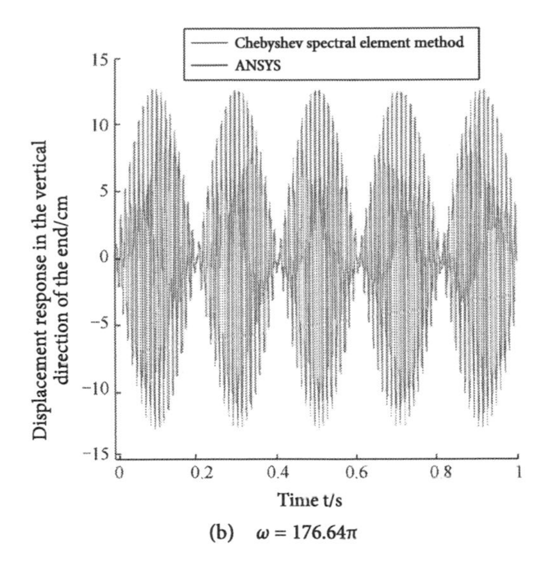

(b) $\omega = 176.64\pi$

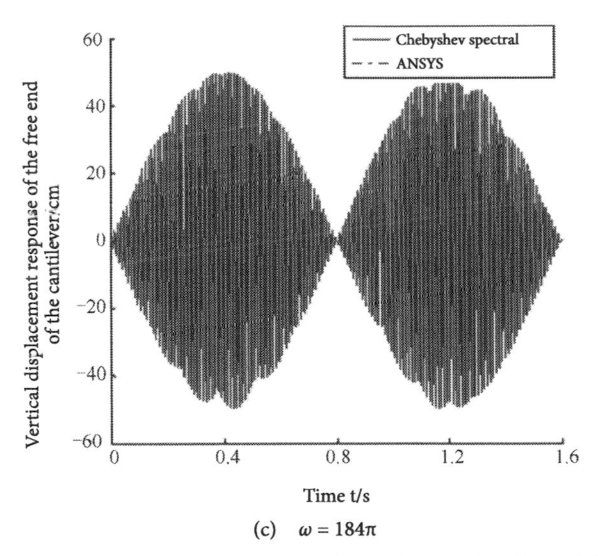

(c) $\omega = 184\pi$

Figure 2-11 Chebyshev spectral element method for solving the vibration problem of a cantilever beam under sinusoidal load (continued)

It can be seen from Figure 2-11 that for the vibration of the cantilever beam under the action of sinusoidal load, the results obtained by applying the Chebyshev spectral element method are in good agreement with the results of ANSYS.

2.5.2 Clustering element spectral element method

To verify the advantages of the clustering element method, four different examples are analyzed and compared with the isometric element method in this chapter.

1. *Standard form*

$$\begin{cases} \dfrac{dx}{dt}+0.6x=10\exp\left[\dfrac{-(t-0.5)^2}{\varepsilon}\right], \ 0\leqslant t\leqslant 1 \\ x(0)=0.5 \end{cases} \tag{2.57}$$

For the sudden load, the conventional isometric element method cannot achieve good convergence. For the impact load shown in Figure 2-12, no matter how complex the system, its dynamic response equation can finally be converted into the standard form shown in Equation (2.57). Among them, at 0.5 s, there is a great impact. The amplitude of the impact can be determined by the coefficient, and the duration of the impact can be changed by adjusting ε, in this case, $\varepsilon = 0.0001125$. The analytic solution can be obtained by integrating the factor method and borrowing the erf function in MATLAB:

$$x(t)=\exp(-0.6t)\left[A_1\text{erf}(A_2-A_3t)+A_4\right]$$

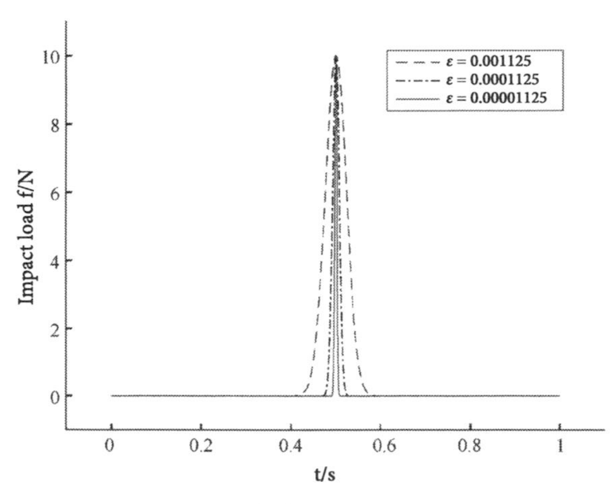

Figure 2-12 Impact load

In the above formula, $A_1 = -5\sqrt{\pi\varepsilon}\exp\dfrac{\left(\dfrac{1+0.6\varepsilon}{2}\right)^2-\dfrac{1}{4}}{\varepsilon}$, $A_2 = \dfrac{1+0.6\varepsilon}{2\sqrt{\varepsilon}}$, $A_3 = \dfrac{1}{\sqrt{\varepsilon}}$, $A_4 = \dfrac{1}{2} - A_1$, $\operatorname{erf}(z) = \dfrac{2}{\sqrt{\pi}}\int_0^z \exp(-w^2)\mathrm{d}w$.

Figure 2-13 shows the calculation results of the equidistant element spectral element method. The error is very large, especially at the sudden load change. When $N_{el} = 30$, the error is extremely large; when $N_{el} = 100$, the error is reduced a lot, but it is still bigger. Figure 2-14 shows the calculation results of the clustered element spectral element method. This method can eliminate errors under the premise that the numbers of elements and the times of interpolations are equal. Figure 2-15 and Figure 2-16 can better illustrate the advantages of the clustered unit spectral element method. Figure 2-15 compares the equidistant element spectral element method with the aggregate element spectral element method. When the number of elements is 50, when the number of element interpolation is 18, the error of the former is 0.02506 and the error of the latter is 1.382×10^{-10}, which has obvious advantages. Figure 2-16 compares the aggregation unit spectral element method with the equidistant unit spectral element method and finds that when the number of units is 80, when the unit interpolation times are 5, 10, and 18, the errors of the aggregation unit spectral element method are 3×10^{-9}, 3.628×10^{-12}, 2.668×10^{-12}, and when the number of element interpolation is 18, the error is 0.01124. It can be seen that the clustered element spectral element method has great advantages in solving the impact load response.

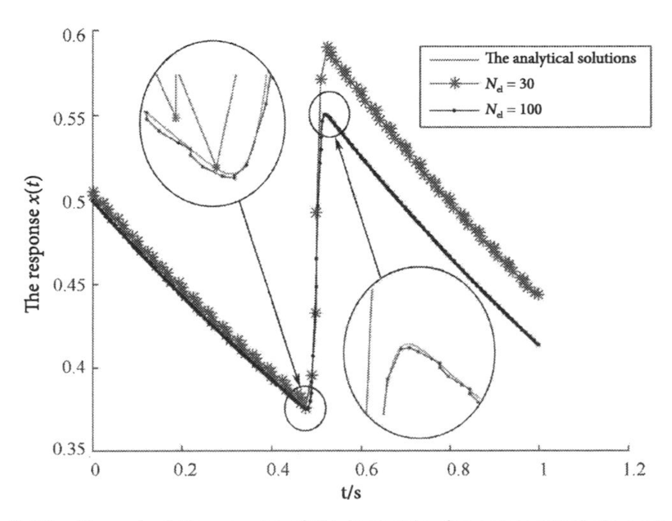

Figure 2-13 The calculation results of the isometric element spectral element method

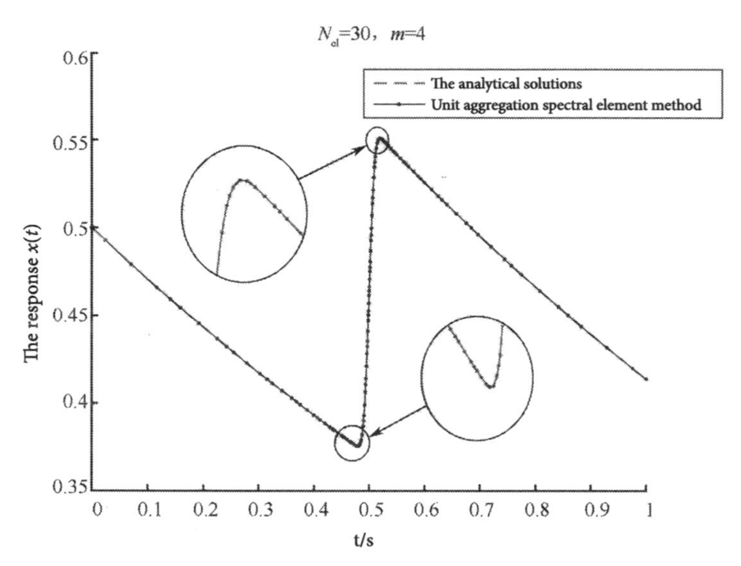

Figure 2-14 The calculation results of the clustered unit spectral element method

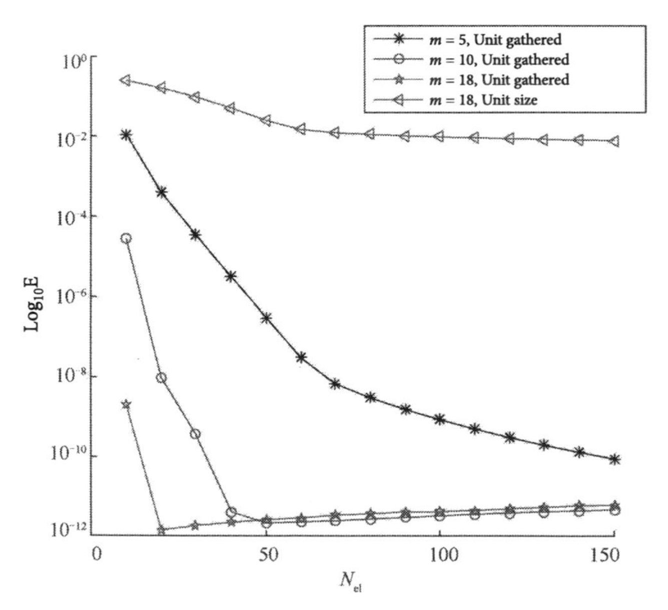

Figure 2-15 Comparison of the aggregated unit spectral element method and the equidistant unit spectral element method (the number of units is 50)

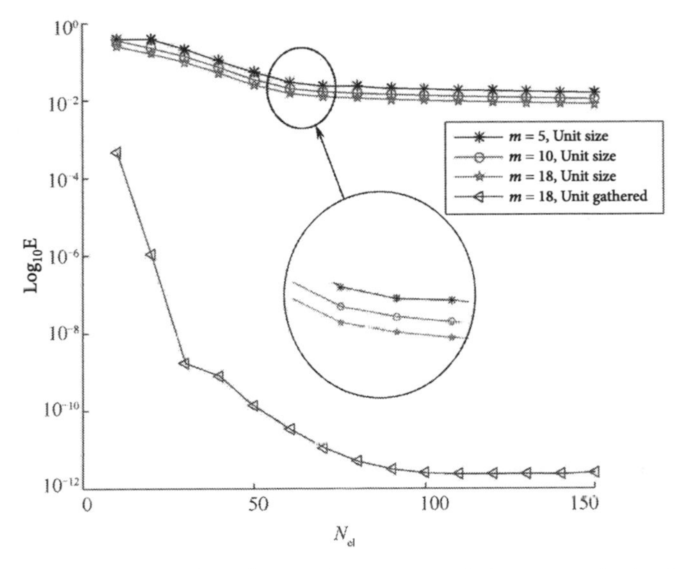

Figure 2-16 Comparison of equidistant element spectral element method and aggregate element spectral element method (the number of units is 80)

2. *Linear single degree of freedom system*

Figure 2-17 shows a linear single degree of freedom system. Among them, the fixed mass $m = 1$ kg, the spring stiffness coefficient is $k = 0.9$, and the damper coefficient is $c = 0.9$. When $t = 0$, the system hits a fixed obstruction at a speed of $v = 1$ m/s, and the impact load $f(t) = 10\exp\left[-(t-6)^2/\varepsilon\right]$, $\varepsilon = 0.01125$. The equation of motion of the system is

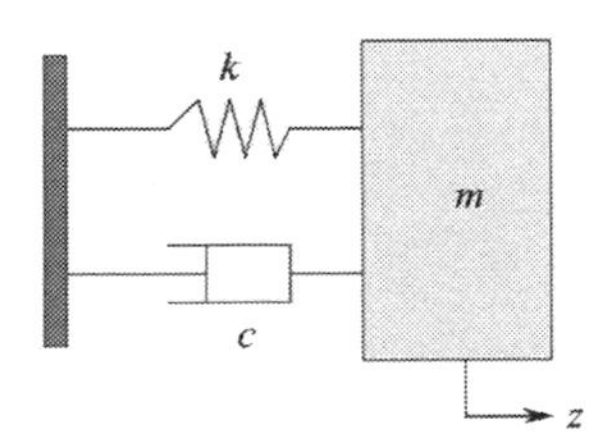

Figure 2-17 Linear single degree of freedom system

$$\ddot{x}(t)+c\dot{x}(t)+kx(t)=f(t) \tag{2.58}$$

The analytic solution of free vibration of Equation (2.58) is

$$z(k,c,m,t)=\begin{cases}\dfrac{\exp - t\dfrac{c}{2m}}{\sqrt{\dfrac{k}{m}-\left(\dfrac{c}{2m}\right)^2}}\sin\left[t\sqrt{\dfrac{k}{m}-\left(\dfrac{c}{2m}\right)^2}\right], \text{if } 0\leqslant\dfrac{c/(2m)}{\sqrt{k/m}}<1\\ t\exp - t\sqrt{\dfrac{k}{m}}, \text{if } \dfrac{c/(2m)}{\sqrt{k/m}}=1\\ \dfrac{\exp - t\dfrac{c}{2m}}{\sqrt{\dfrac{k}{m}-\left(\dfrac{c}{2m}\right)^2}}\left[\exp t\sqrt{\left(\dfrac{c}{2m}\right)^2-\dfrac{k}{m}}-\exp - t\sqrt{\left(\dfrac{c}{2m}\right)^2-\dfrac{k}{m}}\right], \text{if } \dfrac{c/(2m)}{\sqrt{k/m}}>1\end{cases} \tag{2.59}$$

Although the linear single-degree of freedom system has both initial velocity and impact load, the center of impact load is at 6 s, and its width is controlled by $\varepsilon = 0.01125$, which is very narrow. Therefore, the system still belongs to free vibration in the vicinity of 4 s. Figure 2-18 shows the dynamic displacement response of the linear single-degree-of-freedom system. As can be seen from the figure, when the number of elements is 12 and the number of element interpolation is 12, the clustering element spectral element method is more accurate than the isometric element spectral element method.

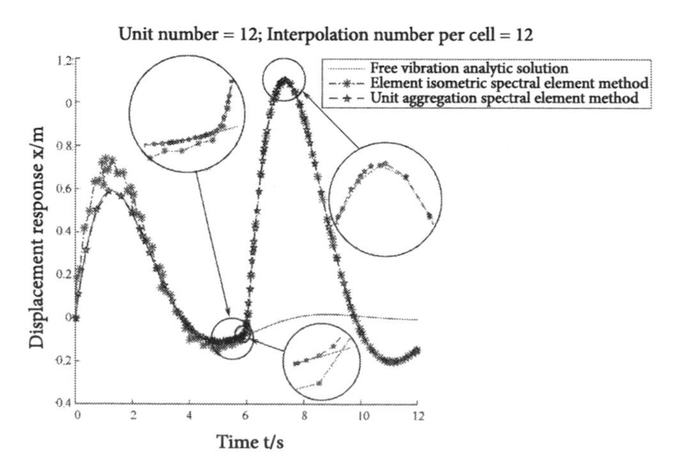

Figure 2-18 The dynamic displacement response of a linear single degree of freedom system

3. *Truss structure*

The 124-bar truss structure contains 49 hinges and 94 degrees of freedom (see Figure 2-19). Its elastic modulus $E = 207$ *GPa*, Poisson's ratio $v = 0.3$, density $\rho = 7850$ kg/m³, and the cross-sectional area of the rod is 0.645×10^{-4} m². The same dynamic load is applied in the positive X direction of nodes 1, 20, 19, 18, 17, 16 and 15, and the negative y-direction of nodes 1, 2, 3, 4, and 5. Dynamic load $f(t) = b\exp\left[-(t-0.2)^2/\varepsilon\right]$, where $b = 1000$, $\varepsilon = 0.0001125$.

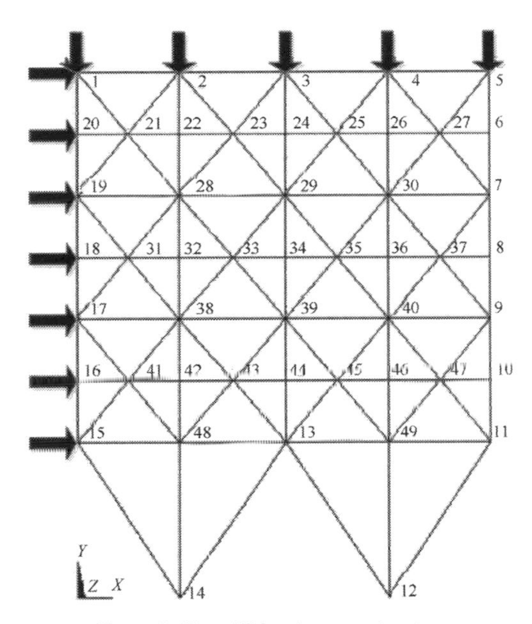

Figure 2-19 124 pole truss structure

From the engineering point of view, nodes 1, 2, 3, 4, 5, 15, 16, 17, 18, 19, and 20 are considered as key locations (called key locations), which are also locations where impact loads are applied. Under the action of impact load, the dynamic displacement response of the key position node of the 124-bar truss structure in the X direction is shown in Figure 2-20. Consider the displacement of the node in the 1X direction separately and enlarge it locally, as shown in Figure 2-21.

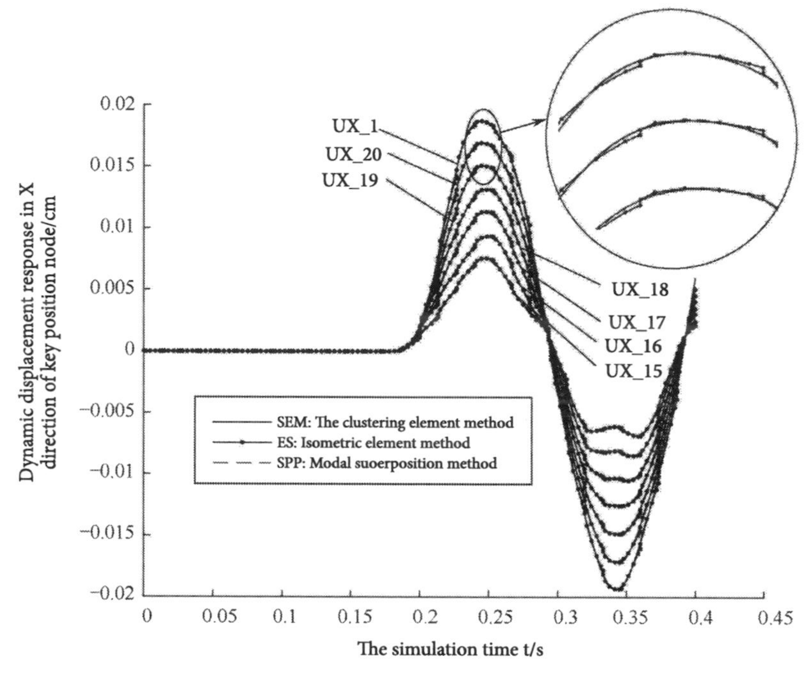

Figure 2-20 Dynamic displacement response in X-direction of key nodes in 124 bar truss structure

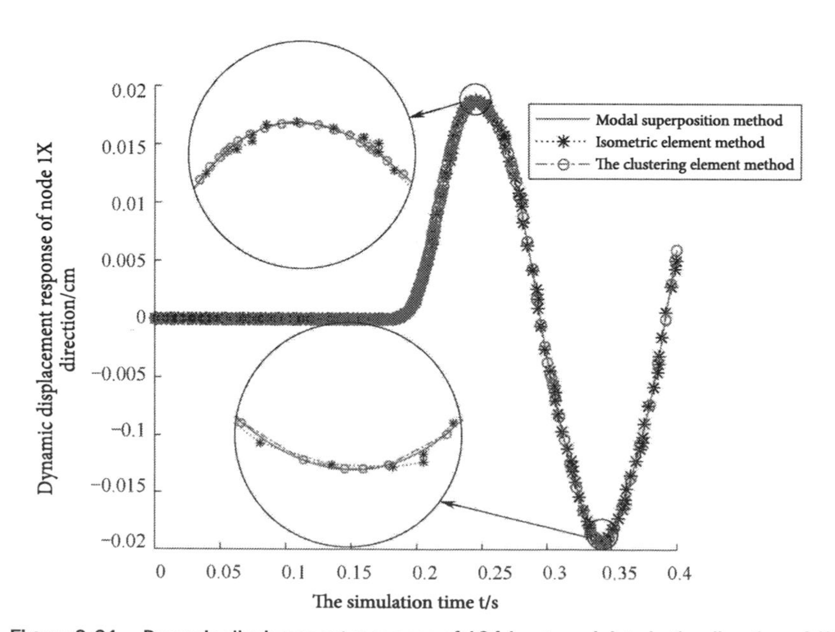

Figure 2-21 Dynamic displacement response of 124 bar truss joints in the direction of 1X

It can be seen that the results of the equidistant element spectral element method deviate from the exact value at some points, while the aggregated element spectral element method does not. Each node in the key position has a similar phenomenon, which is limited in space and not listed in full.

4. Connecting rod small end

The small end of the connecting rod and its connected body are shown in Figure 2-22. It adopts PLANE42 cell with fixed right end, 111 nodes, and 212 degrees of freedom. Horizontally to the right impact load in the middle of the small head inner circle of the connecting rod, $f(t) = b\exp\left[(-(t-0.1)^2/\varepsilon)\right]$, $b = 10^6$, $\varepsilon = 0.0001125$. The elastic modulus of the connecting rod material $E = 207GPa$, Poisson's ratio $\nu = 0.3$, density $\rho = 7850$ kg/m^3.

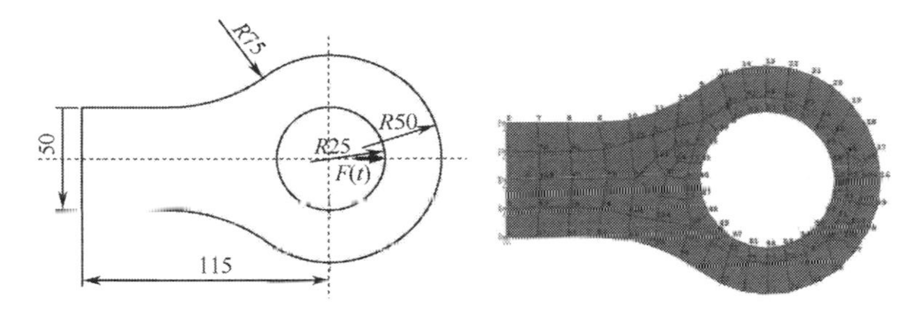

Figure 2-22 The small end of the connecting rod and its connecting rod body part

The dynamic displacement response of nodes 15, 42, 49, and 93 in the X direction was investigated by taking the rounded corner of the small head of the connecting rod as the key position. In Figure 2-23, if the results obtained by the modal superposition method are regarded as accurate solutions, it can be seen that under the same number of elements, size, and interpolation times, the results obtained by the clustering element spectral element method are more accurate and consistent with those obtained by the modal superposition method. All the while, the results obtained by the isometric element spectral element method have large errors. As the displacement response itself is small, Figure 2-23 is also partially enlarged.

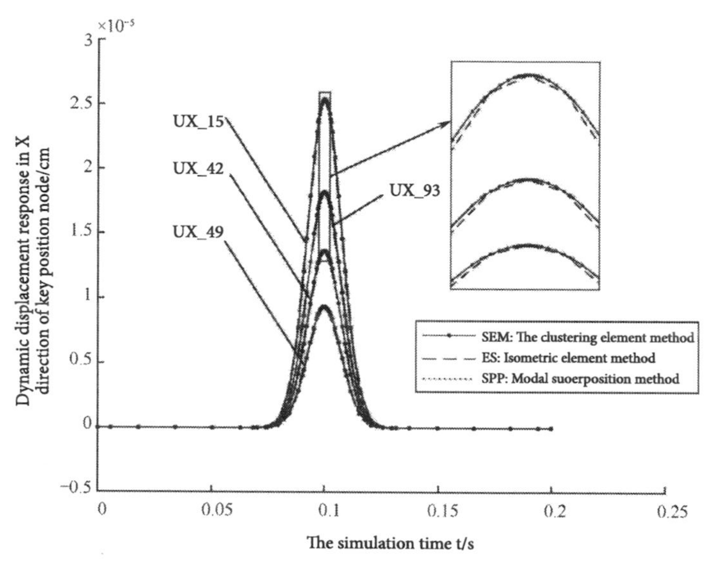

Figure 2-23 X-direction dynamic displacement response of the four key position nodes of the small head of the connecting rod

2.5.3 Nonlinear vibration analysis

1. Nonlinear vibration equation of the Duffing type

The nonlinear vibration equation of Duffing type can be written as

$$\ddot{u}+u+\varepsilon u^3=F\sin(\omega t) \tag{2.60}$$

In the formula, ε and F are given constants; ω is the frequency of the external load, which is also a constant.

(1) When $\varepsilon=-1/6$, $F=0$, and $\omega=0.7$, the initial conditions were

$$\begin{cases} u(0)=0 \\ \dot{u}(0)=1.62376 \end{cases}$$

The approximate analytic solution of Equation (2.60) is

$$u(t)\cong 2.058\sin(0.7t)+0.0816\sin(2.1t)+ 0.00337\sin(3.5t) \tag{2.61}$$

At this point, the response to the Duffing nonlinear vibration problem and the Newton-Raphson iteration process is shown in Figure 2-24 and Figure 2-25, respectively.

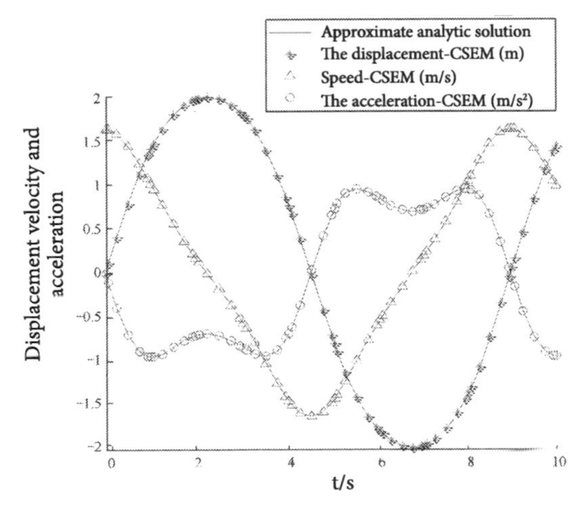

Figure 2-24 Response of Duffing-type nonlinear vibration problem (the first initial condition)

(2) When $\varepsilon = -1/6$, $F = 2$ and $\omega = 1$, the initial conditions were

$$\begin{cases} u(0) = 0 \\ \dot{u}(0) = -2.7676 \end{cases}$$

The approximate analytic solution of Equation (2.60) is

$$u(t) \cong -2.5425\sin t - 0.07139\sin(3t) - 0.00219\sin(5t) \tag{2.62}$$

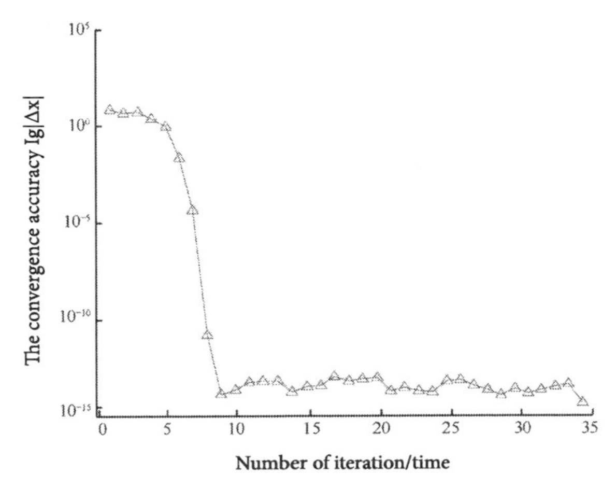

Figure 2-25 Newton-Raphson iterative process of Duffing-type nonlinear vibration problem (The first initial condition)

At this time, the response of the Duffing-type nonlinear vibration problem and the Newton-Raphson iteration process are shown in Figure 2-26 and Figure 2-27, respectively.

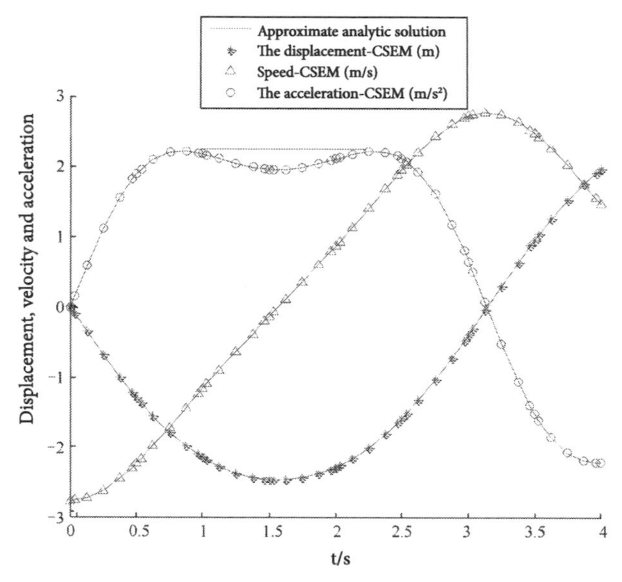

Figure 2-26 Duffing type nonlinear vibration problem response (the second initial condition)

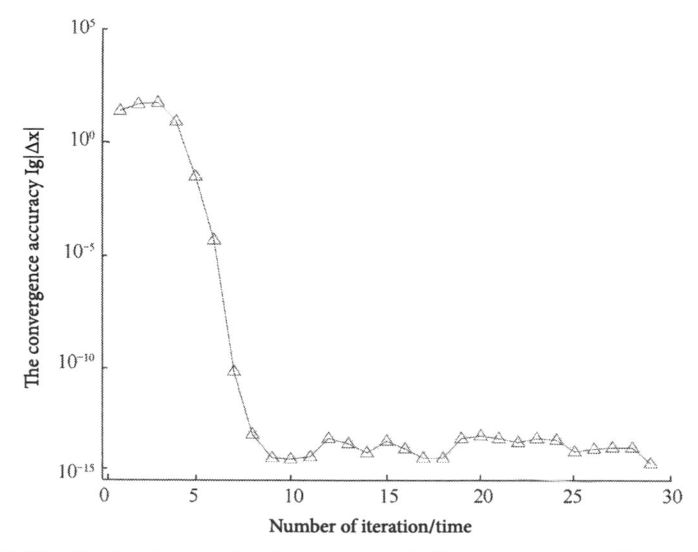

Figure 2-27 Newton-Raphson iterative process of Duffing-type nonlinear vibration problem (The second initial condition)

As can be seen from Figure 2-24 and Figure 2-26, the results obtained by this method in this chapter are very consistent with the approximate exact solution. It can be seen from Figure 2-25 and Figure 2-27 that the Duffing-type nonlinear vibration has a good convergence under two different parameters.

2. *Nonlinear oscillation of a simple pendulum*

The nonlinear vibration equation of a simple pendulum is

$$\ddot{\theta} + \frac{g}{l}\sin\theta = 0 \tag{2.63}$$

In the formula, g is the acceleration due to gravity; l is the pendulum length; θ is the pendulum angle. The initial conditions are $\theta(0) = \theta_0$, $\theta(0) = \dot{\theta}_0$.

When the number of elements is 10 and the number of interpolation times is 6, the nonlinear equations are obtained by the Galerkin discrete scheme, and then solved by the Newton-Raphson method, the pendulum Angle, angular velocity, and angular acceleration when the initial pendulum Angle $\theta_0 = \pi/5$ can be obtained, as shown in Figure 2-28. This is in good agreement with the calculation results of the ODE45 solver. Figure 2-29 shows the Newton-Raphson iterative process of the nonlinear vibration of a single pendulum, and it can be seen that the nonlinear oscillation has a good convergence.

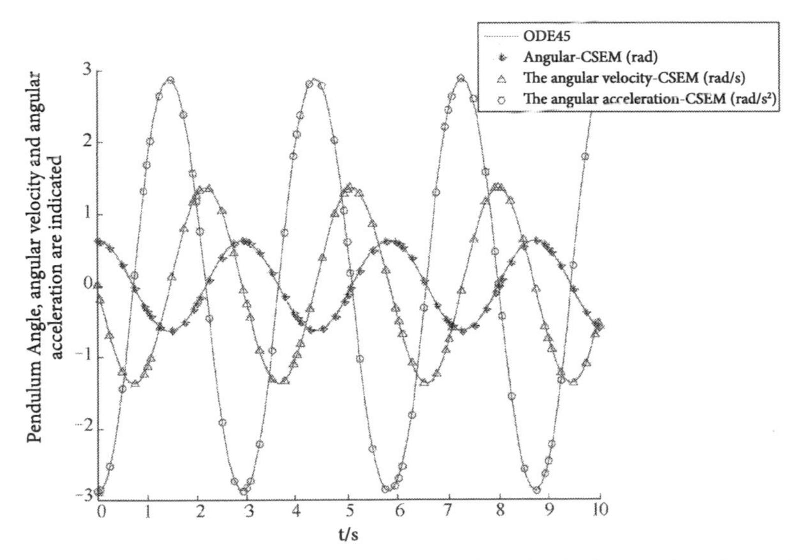

Figure 2-28 The response of the nonlinear vibration of a simple pendulum $\theta_0 = \pi/5$

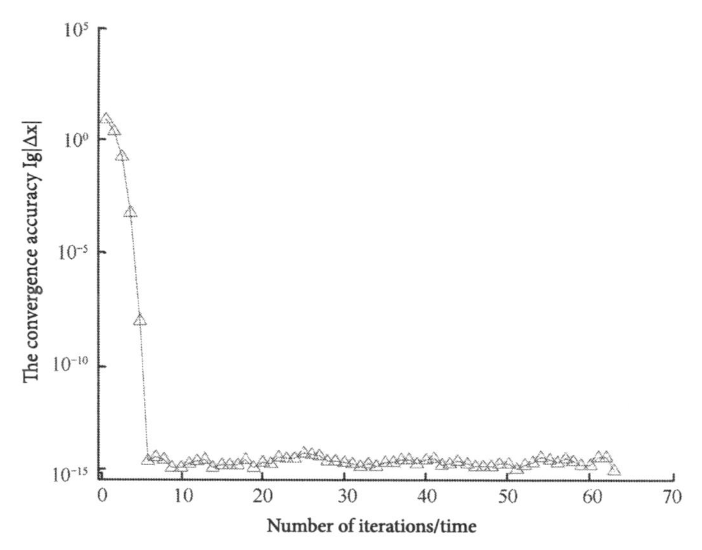

Figure 2-29 Newton-Raphson iteration process of the nonlinear vibration of a simple pendulum

After solving the pendulum Angle response $\theta(t)$, two time points t_i and t_j can be obtained, which satisfy $\theta(t_i) > 0$, $\theta(t_j) < 0$ and $t_i < t_j$. The Lagrange interpolation of the center of gravity is carried out on the interval $[t_i, t_j]$, and then $\theta(t) = 0$ is solved by dichotomy, and $t_{\theta=0}$ can be obtained. Then, the period $T = 4t_{\theta=0}$ of the nonlinear oscillation of a single pendulum with an initial pendulum Angle. Compared with the period $T_0 = 2\pi\sqrt{l/g}$ of the linear vibration of A single pendulum, as shown in Table 2.5, and compared with the exact solution, the second-order perturbation solution, and the DQ method.

According to the data in Table 2.5, when the initial pendulum Angle $\theta_0 < 135°$, the spectral element method can obtain the maximum absolute error of 0.0013, while the maximum absolute error of second-order perturbation solution is 0.1042, and the maximum absolute error of DQ method is 0.0016. When $\theta_0 = 150°$, the maximum absolute error obtained by the spectral element method is 0.0016, the maximum absolute error obtained by the second-order perturbation solution is 0.1585, and the maximum absolute error obtained by the DQ method is 0.0125.

Table 2.5 Ratio of initial pendulum Angle and natural frequency of nonlinear vibration of a single pendulum

The initial angular	The exact solution [109]	The spectral element method	Second-order perturbed solution [110]	DQ method [111]
5°	0.9995	0.999524054146334	0.9995	0.9994
15°	0.9957	0.995717836609230	0.9957	0.9957
30°	0.9829	0.982889082896578	0.9829	0.9829
60°	0.9318	0.931808391622448	0.9335	0.9319
90°	0.8472	0.847213084794106	0.8620	0.8472
120°	0.7285	0.728395515523455	0.7895	0.7283
135°	0.6558	0.654472710779005	0.7600	0.6542
150°	0.5791	0.567471276593891	0.7376	0.5666

2.6 Summary of this chapter

In this chapter, the exact element interpolation differential matrix is obtained by using the unknown approximate element function of Lagrange interpolation, and the global interpolation differential matrix is obtained by the finite element node sharing property. Chebyshev spectral element method can obtain high precision for vibration problems with arbitrary loads, especially when the simulation time is long, it can show its superiority, and it can overcome the shortcoming that the matching method is unable to get available solutions for vibration problems with long simulation time. Chebyshev spectral element method can also obtain satisfactory solutions not only for one-dimensional vibration problems under arbitrary loads, but also for continuous vibration problems under arbitrary loads, which provides a reference for further solving dynamic problems and their optimal design.

According to the characteristics of the impact load and combining with the advantages of high precision of spectral element method, the ratio of the size of the load center unit and the size of the two sides of the unit is set to adapt to the load mutation. This is done so that the unit's dispersion can adapt to the change of the impact load. Under the premise of the same number of elements, size, and interpolation times, the clustering element spectral element method is more accurate than the isometric element spectral element method in solving the dynamic response problem of the impact load. From the standard form of dynamic problems to linear single-degree of freedom systems and 124 bar plane trusses, the results of dynamic response analysis of impact load for the small-head problem of the connecting rod show the feasibility and effectiveness of the spectral element method. This provides a reference for the further study of the dynamic response of the structure under impact load.

By obtaining the nonlinear algebraic equations and combining the Newton-Raphson method, the displacement and velocity of the nonlinear vibration problem can be obtained simultaneously, and then the acceleration can be obtained through the relationship between the added velocity and displacement and velocity in the differential equation. For the problem of nonlinear single pendulum vibration, after the angular displacement is calculated, the angular frequency at different initial pendulum angles can be precisely calculated by combining the dichotomy method, and it can compared with other methods, thus showing that the spectral element method has the highest accuracy.

CHAPTER 3

Dynamic Response Optimization Method Based on Time Spectrum Element Method

For a mechanical structure that can establish an accurate physical model, under the action of dynamic load, and can make the dynamic performance index of the machine reach the extreme value, we need to establish the dynamic response equation, and then optimize the dynamic response. Of course, parallelization can play a certain role in dynamic response optimization, but it is only considered from the perspective of reducing time consumption. However, in the process of dynamic response optimization solution, the accuracy of the simulation model brings difficulties to the dynamic response optimization, so it is necessary to study a high-precision and high-efficiency technique to solve the dynamic response for such problems.

The finite difference method is dominant in the numerical calculation method of time-dependent response. This method starts from the initial conditions and uses small time steps to calculate time-related second-order differential equations or other equations, until the convergence requirements are met. The size of the time step determines the stability and accuracy of the calculation, and also affects the calculation efficiency. If the period response is calculated after multiple transient periods and reaching a steady-state cycle, the calculation efficiency will be significantly improved. However, when the system is excited by pulses, the frequency-based method has insufficient accuracy in solving the problem of rapid response changes. For the dynamic system stimulated by a pulse, the spectral element method of h-fine scheme is adopted near the semicontinuous, and the differential equation or equations which are transformed into algebraic equations serve to achieve the spectral convergence accuracy without increasing the number of elements. [44]

Three schemes were proposed that use design variables and displacement responses, speed responses, and acceleration responses as optimization variables. [112] The finite difference method is used to approximate time-related constraints and the differential equation of motion is treated as an equation constraint so that the gradient expression of the optimized variable is given in an explicit form. This is done so that the gradient information can be efficiently obtained. However, the turning of a small-variable problem into a high-dimensional problem is not only a difficult one to solve, but also difficult to converge. For the smallest problem in the maximum dynamic response, since the maximum response is oscillating during the iteration process, if the objective function is directly processed, it is difficult to converge. [113] Therefore, a global maximum response coefficient can be given in advance, and in the optimization process, the local dynamic response maximum larger than this global maximum response coefficient should be minimized.

The biggest characteristic of the finite element method is that it transforms differential equations or other equations into algebraic equations [114]. In this way, more explicit relations of response to design variables can be obtained, and the sensitivity of response can be effectively calculated. In this book, we use the temporal spectral element method to discretize differential equations or differential equations in the time domain. In 1984, Patera proposed the spectral element method, [42] which not only has the flexibility of processing boundary and structure of finite element method, but also the fast convergence of the spectral method. In a unit, the spectral element method discretizes time into grid points corresponding to the zero points of the GLL polynomial. It also performs Lagrange interpolation on these points and then solves the differential equation or other equations over the entire time interval to obtain the transient response and steady-state response. When solving the impulse excitation response, the spectral element method is used to perform time discretization over the entire time interval, and to reflect its local flexibility [44] by increasing the number of interpolations near the excitation mutation, or reducing the unit size.

The structural optimization of withstanding transient loads has been studied since the 1970 s. In 2006, Kang et al. [37] first took dynamic response optimization as a branch of optimization. In its optimization, it is necessary to calculate the time response for each iteration of the design variables, and at the same time, the constraints must be met in the entire time interval. The method of processing time constraints is that one is to satisfy the constraint only at the global maximum of the response, while the other is to satisfy the constraint at a smaller time step. This is done so that constraint failures are unlikely to occur at the intermediate point, making the method more stable. The quasi-static method applies to this method to make multiple equal static loads satisfy constraints. [115] In these two methods, the number of constraints is greatly increased. Since the sensitivity of these constraints must be calculated during the optimization iteration process, the cost of such optimization is also increasing. There is also a more effective method of dealing with time constraints, that is, it only handles constraints at the local extreme points of the response, which reduces the number of constraints and reduces the number of times the sensitivity of the constraints to the design variables that is calculated. [116]

This chapter uses the time spectrum element method to effectively optimize the excitation dynamic system and compares the optimization cost of implementing constraints at the GLL point and the local extreme points of the response.

3.1 Dynamic Response Optimization Model of Mechanical Structure

The objective function of the mechanical structure dynamic response optimization design problem is described here. First, select design variable vector $X(x_1, x_2, \cdots x_k)$, to satisfy the constraints of instantaneous dynamic performance (Amplitude or relative displacement limit, dynamic stress, dynamic strain failure limit, etc) and the allowable range of design variables within a given time interval [0,T] under the action of transient load. Second, make the maximum dynamic response (Displacement, velocity, acceleration) of some key position coordinates of the system reach their best under a certain meaning or criterion. Its mathematical model can be expressed as:

$$\begin{cases} \boldsymbol{X} = \left[x_1, x_2, \cdots, x_k\right]^{\mathrm{T}} \\ z(t) = \left[z(t), \dot{z}(t), \ddot{z}(t)\right]^{\mathrm{T}} \\ \dot{z}(t) = p\left[\boldsymbol{X}, t, z(t), \boldsymbol{F}(t)\right] \\ z(t_0) = z_0, \dot{z}(t_0) = \dot{z}_0 \\ \boldsymbol{J}(\boldsymbol{X}, z, t) \\ h_i\left[\boldsymbol{X}, t, z(t)\right] = 0, i = 1, 2, \cdots, m \\ g_j\left[\boldsymbol{X}, t, z(t)\right] \leqslant 0, j = 1, 2, \cdots, n \\ \forall t \in \left[0, T\right] \end{cases} \tag{3.1}$$

In this formula, $\boldsymbol{X}$ is the design variable vector of the system, which is composed of geometric and physical parameters of the system; $z(t)$ is the state variable vector of the system, which is composed of the generalized coordinates of the motion state of the system, such as displacement, velocity, acceleration and so on. It must satisfy the motion equation of the mechanical system, that is the mathematical description of dynamic characteristics, $\dot{z}(t) = \boldsymbol{p}\left[\boldsymbol{X}, t, z(t), \boldsymbol{F}(t)\right]$; The $\boldsymbol{J}$ is the objective function, and the h_i, g_i represents the equality constraint and the inequality constraint respectively, which must be satisfied at all time points $\forall t \in [0, T]$. Generally established equations of motion are second-order differential equations, but the equations of motion in Equation (3.1) are first-order. We can use variable substitution to transform the second-order differential equations, or second-order differential equations into First-order differential equations.

There are two ways to deal with time-related constraints (see [117]). One is to satisfy the constraint at all GLL points, that is, the GLL point method and the other is to satisfy the constraint at the absolute value extreme point of each unit, that is, the key point method. The first method requires the distance between GLL points to be as small as possible so that the possibility that the absolute extreme points between GLLs that cannot be satisfied are reduced. Therefore, the number of constraints is relatively large, and the solution cost of the spectral element method is also relatively high. The second method can solve the differential equation of motion by using fewer spectral elements which meet the requirement of accuracy. In each unit, it finds the absolute value extreme point of the unit by performing a one-dimensional search on the high-order L function. Of course, the absolute value extreme points of the element are oscillating. Therefore, at each iteration step, the absolute extreme value points of the element must be recalculated.

3.2 Dynamic response optimization method

In a variety of mechanical systems (such as a shock absorber, weapon rear-seat mechanism, aircraft landing gear, automobile suspension system, etc.) The main mass of the system may have a large vibration before the system reaches a steady state, and the system may be destroyed when the frequency of the excitation force is close to the natural frequency of the system. Therefore, the instantaneous dynamic response of the excitation must be constrained. For this kind of problem, the optimization of the physical model can be used to solve it.

3.3 Optimal design of linear single degree of freedom system

The linear single-degree-of-freedom system shown in Figure 3-1 has a mass with a fixed mass $m = 1$ kg, and the design variables k and c represent the spring stiffness coefficient and damping coefficient, respectively. When $t = 0$, the system hits a fixed obstruction at a speed of $v = 1$ m/s. Design k and c so that at the time [0, 12 s], the acceleration of the mass is minimized and the maximum displacement response is not greater than 1m.

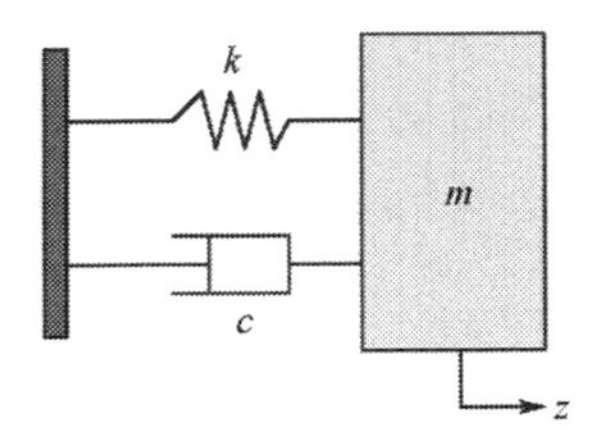

Figure 3-1 Linear Single-DOF Systems

The system motion equation is:

$$\ddot{z}(t)+c\dot{z}(t)+kz(t)=0 \tag{3.2}$$

The analytical solution of Equation (3.2) is:

$$z(k,c,m,t)=$$

$$\begin{cases} \dfrac{\exp -t\dfrac{c}{2m}}{\sqrt{\dfrac{k}{m}-\left(\dfrac{c}{2m}\right)^2}}\sin\left(t\sqrt{\dfrac{k}{m}-\left(\dfrac{c}{2m}\right)^2}\right), & \text{if } 0\leqslant\dfrac{c/(2m)}{\sqrt{k/m}}<1 \\ t\exp -t\sqrt{\dfrac{k}{m}}, & \text{if } \dfrac{c/(2m)}{\sqrt{k/m}}>1 \\ \dfrac{\exp -t\dfrac{c}{2m}}{\sqrt{\dfrac{k}{m}-\left(\dfrac{c}{2m}\right)^2}}\left[\exp t\sqrt{\left(\dfrac{c}{2m}\right)^2-\dfrac{k}{m}}-\exp -t\sqrt{\left(\dfrac{c}{2m}\right)^2-\dfrac{k}{m}}\right], & \text{if } \dfrac{c/(2m)}{\sqrt{k/m}}>1 \end{cases} \tag{3.3}$$

When dealing with the objective function, the artificial design variable b is introduced, then the optimization model of the linear single degree of a freedom system problem is:

$$\begin{cases} \min b \\ \ddot{z}(t) + c\dot{z}(t) + kz(t) = 0 \\ \left|c\dot{z}(t) + kz(t)\right| - b \leqslant 0 \\ \left|z(t) - 1 \leqslant 0\right| \end{cases} \tag{3.4}$$

The optimization model formula (3.4) is solved by the analytical method and the spectral element method respectively, and the analytical solution is shown in Figure 3-2. The solid circle in Figure 3-2 is the best point. The results of the spectral element method are shown in Table 3.1–Table 3.4 and Figure 3-3 and Figure 3-4. Among them, Table 3.1, Table 3.2, and Figure 3-3 are the results obtained by using the GLL point method; Table 3.3, Table 3.4, and Figure 3-4 are the results obtained by using the key point method.

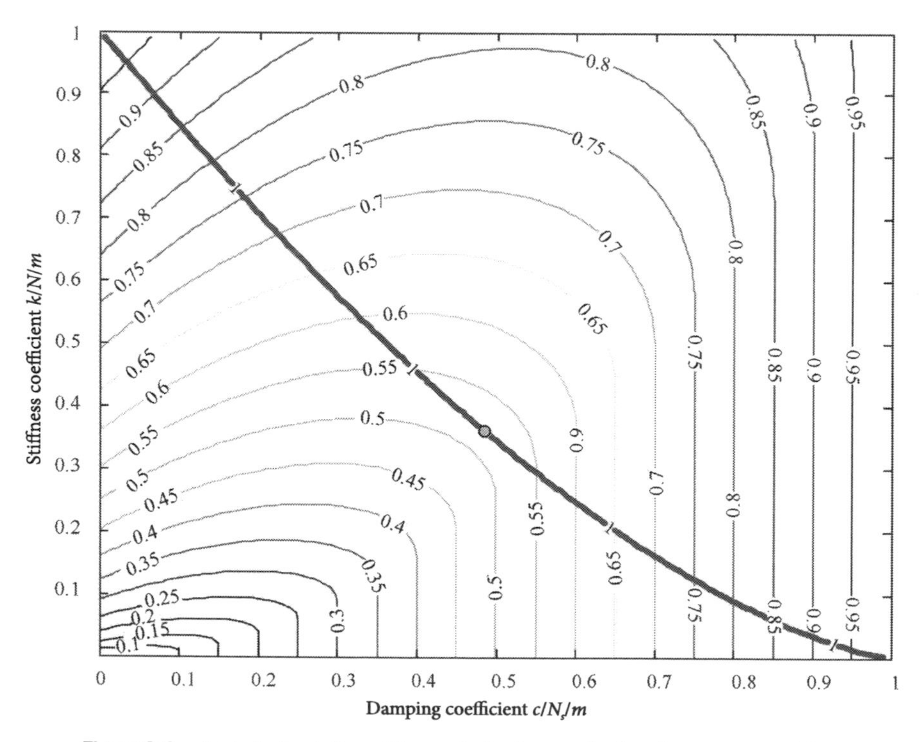

Figure 3-2 Analytical solution of linear single degree of a freedom system problem

Table 3.1 The best advantage of the GLL point method numerical test (single degree of freedom)

Number of interpolation points m		Number of units N_{el}									
		10	20	60	100	200	300	400	500	600	800
3	c	0.4684655	0.487330	0.484996	0.484435	0.4840003	0.485851	0.485281	0.4849336	-4.847042×101	0.4851674
	k	0.3746571	0.355675	0.360456	0.361225	0.3617721	0.359815	0.360431	0.3607923	-3.610399×101	0.3605523
	$\lvert a \rvert$	0.5124303	0.518689	0.520384	0.520524	0.5205923	0.520584	0.520593	0.5205929	-5.205972×101	0.5205977
6	c	0.4984423	0.485088	0.484061	0.486063	0.4848087	0.485537	0.485042	0.4855214	0.4851940	0.4852141
	k	0.3464638	0.360561	0.361708	0.359608	0.36093	0.360162	0.360684	0.3601791	0.3603249	0.3605036
	$\lvert a \rvert$	0.5195993	0.520487	0.520592	0.520590	0.520596	0.520596	0.520597	0.5205979	0.5205980	0.5205983
10	c	0.4939818	0.484161	0.485248	0.484568	0.4848257	0.485077	0.485282	0.4849362	0.4851150	0.4851482
	k	0.3510828	0.361601	0.360467	0.361183	0.3609125	0.360646	0.360431	0.3607958	0.3606082	0.3605730
	$\lvert a \rvert$	0.5201820	0.520586	0.520592	0.520598	0.5205981	0.520597	0.520598	0.5205983	0.5205984	0.5205986

NOTE: $\lvert a \rvert$ in the table represents the absolute value of acceleration, the unit is m/s^2, the same below.

Table 3.2 GLL point method CPU time t and iteration number n (single degree of freedom)

Number of interpolation points m		Number of units N_{el}									
		10	20	60	100	200	300	400	500	600	800
3	t	5.468750	3.328125	7.67187	15.4062	75.8750	47.78125	46.10938	105.2656	184.7656	86.12500
	n	8	10	10	12	20	20	15	29	44	13
6	t	3.468750	5.421875	24.3750	70.5312	426.656	186.0469	123.5156	436.4063	221.0781	249.6563
	n	10	11	15	20	27	39	18	53	21	16
10	t	4.218750	14.10938	66.9062	289.421	201.3281	170.7031	633.6250	1872.672	521.7031	585.3890
	n	9	15	18	27	31	15	37	69	16	105

NOTE: The unit of CPU time t is s, the same below.

Table 3.3 The best advantage of the key point method numerical test (single degree of freedom)

Number of interpolation points m		Number of units N_{el}									
		10	20	60	100	200	300	400	500	600	800
3	$c\ k$	0.484269	0.484827	0.485112	0.485137	0.485148	0.485149	0.48515	0.485150	0.485151	0.485151
		0.359551	0.360387	0.360550	0.360563	0.360568	0.360569	0.36057	0.360570	0.360570	0.360570
	$\|a\|$	0.519002	0.520154	0.520547	0.520580	0.520594	0.520597	0.52059	0.520598	0.520598	0.520598
6	$c\ k$	0.485036	0.485122	0.485148	0.485150	0.485151	0.485151	0.48515	0.485151	0.485151	0.485151
		0.360512	0.360555	0.360569	0.360570	0.360570	0.360570	0.36057	0.360570	0.360570	0.360570
	$\|a\|$	0.520447	0.520560	0.520594	0.520597	0.520598	0.520598	0.52059	0.520599	0.520599	0.520599
10	$c\ k$	0.485134	0.485147	0.485151	0.485151	0.485151	0.485151	0.48515	0.485151	0.485151	0.485151
		0.360562	0.360568	0.360570	0.360570	0.360570	0.360570	0.36057	0.360570	0.360570	0.360570
	$\|a\|$	0.520576	0.520593	0.520598	0.520598	0.520599	0.520599	0.52059	0.520599	0.520599	0.520599

Table 3.4 Critical point method CPU time consumption and number of iterations (single degree of freedom)

Number of interpolation points m		Number of units N_{el}									
		10	20	60	100	200	300	400	500	600	800
3	t	16.843750	27.312500	78.484375	143.140625	265.17187	423.0468	538.8593	438.984375	538.046875	781.421875
	n	12	12	12	13	14	15	15	12	13	14
6	t	16.968750	34.156250	114.015625	239.015625	296.95312	686.8906	717.00000	691.421875	853.437500	1122.609375
	n	10	11	13	16	11	18	13	13	14	13
10	t	24.578125	56.500000	163.593750	255.000000	517.01562	1064.531	1170.9843	1760.796875	1495.906250	2910.078125
	n	11	13	13	13	13	17	16	15	14	14

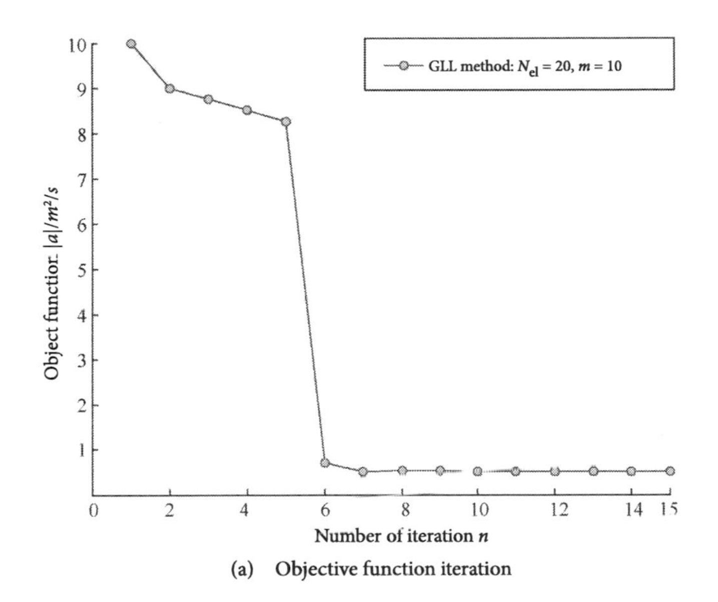

(a) Objective function iteration

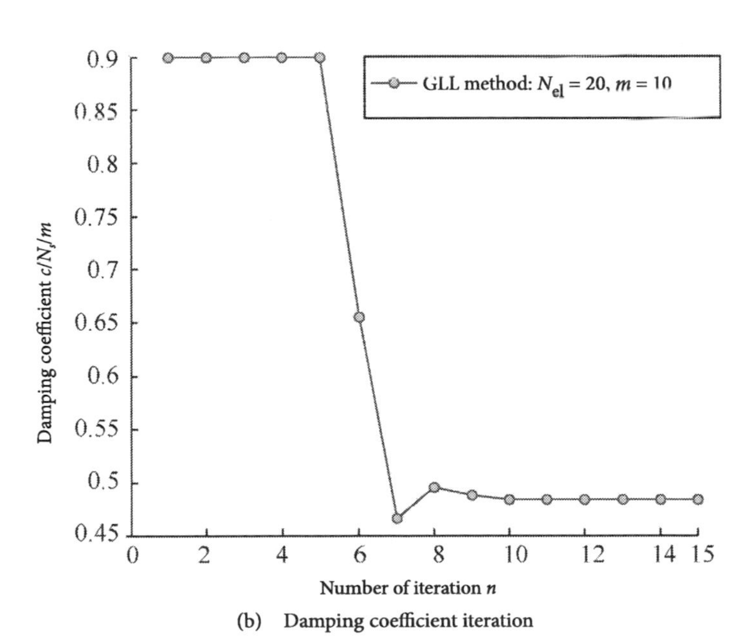

(b) Damping coefficient iteration

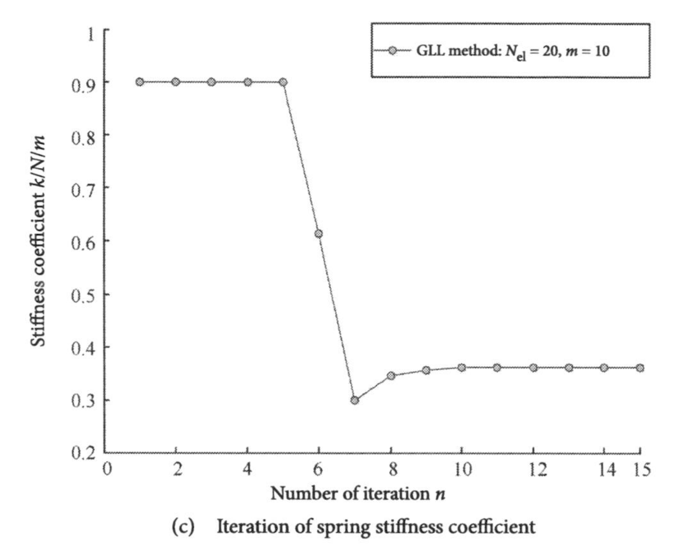

(c) Iteration of spring stiffness coefficient

Figure 3-3 Optimization iteration of GLL point method for linear single degree of freedom system

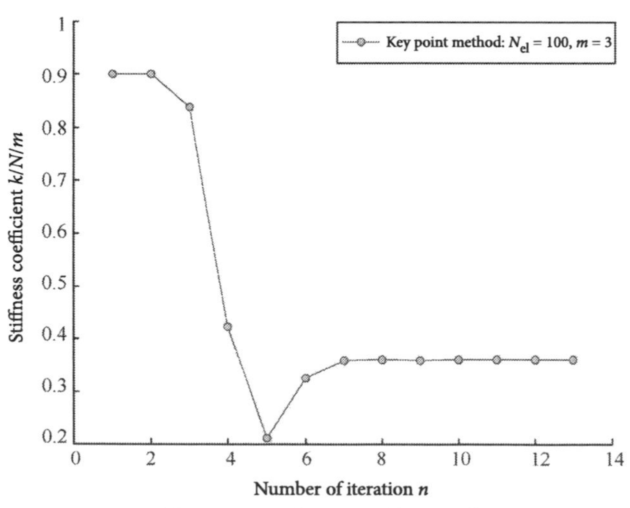

(a) Iteration of spring stiffness coefficient

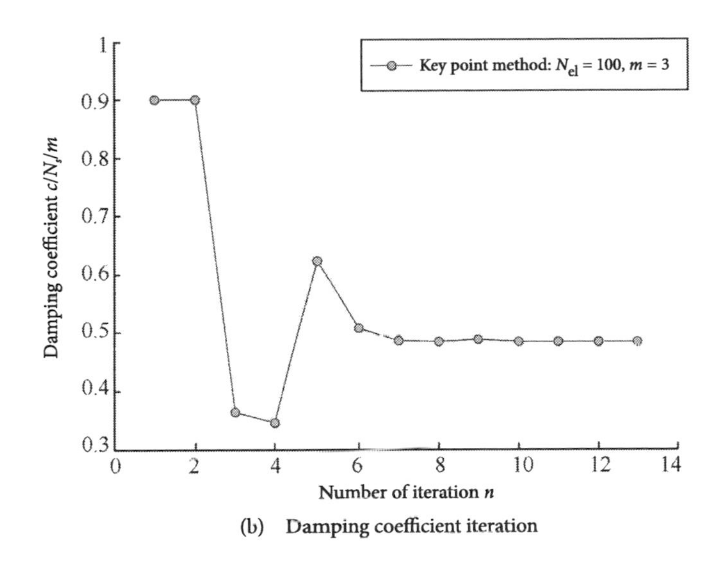

(b) Damping coefficient iteration

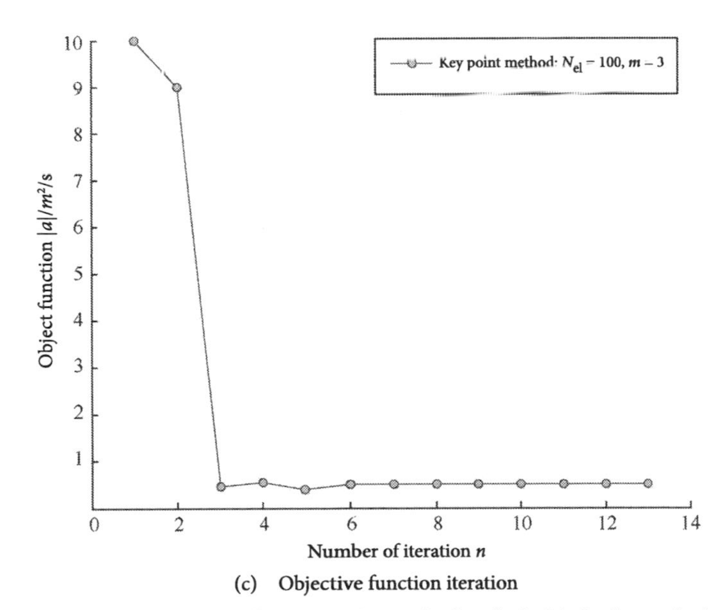

(c) Objective function iteration

Figure 3-4 Optimization iteration of the key point method to deal with the linear single degree of a freedom system problem

When dealing with time constraints by GLL point method and solving the optimal design problem of the linear single-degree-of-freedom system by spectral element method, it can be seen from Table 3.1 and Table 3.2 that the following results can be obtained. Except for the cases where the number of units is 10, 20, 60, the number of interpolation points is 3, the number of units is 10, and the number of interpolation points is 3, 6, and 10, while the exact optimal solution value is not obtained. Other cases both have obtained the optimal solution. Among them, when the number of units is greater than 200, the result is the most accurate. Figure 3-3 shows that the spectral element method using GLL point method to deal with time constraints can quickly converge to the best.

It can be seen from Table 3.1 and Table 3.2 that in the GLL point method test results, all except for the interpolation numbers 3, 6, 10 are corresponding to the element number 10 and the interpolation numbers 3, 6 are corresponding to the element number 20, which all obtained the best points. At the same time, reasonable results show a stepped shape in Table 3.1. It can be seen from Table 3.2 that among these reasonable results, the least time-consuming is 4.218750 s and the most are 585.3890 s. From the analysis in Table 3.3 and Table 3.4, it can be seen that in the key point method test results, except for the number of units of 10 and the number of interpolations of 3, all obtained the best points, while the least time-consuming is 16.968750 s, and the most are 2910.078125 s. Figure 3-4 shows that the spectral element method using the key point method to deal with time constraints can also quickly converge to the optimal point. It can be seen from Figure 3-5 that the spectral element method with the key point method handles time constraints can quickly drop to near the optimal point, while the GLL point method handles the time-constrained spectral element method that is slower, but the final GLL point method is more iterative than the key point method (which is less so). Therefore, in terms of accuracy, the key point method is better than the GLL point method. From the time-consuming point of view, the GLL point method is better than the key point method. Figure 3-6 shows that the optimization effect is very obvious.

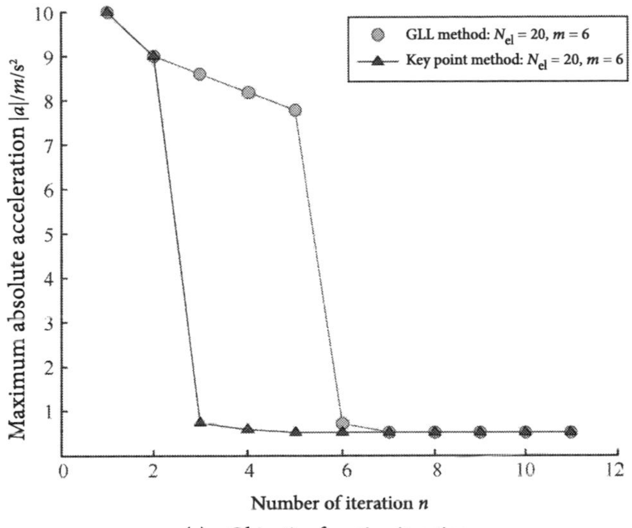

(a) Objective function iteration

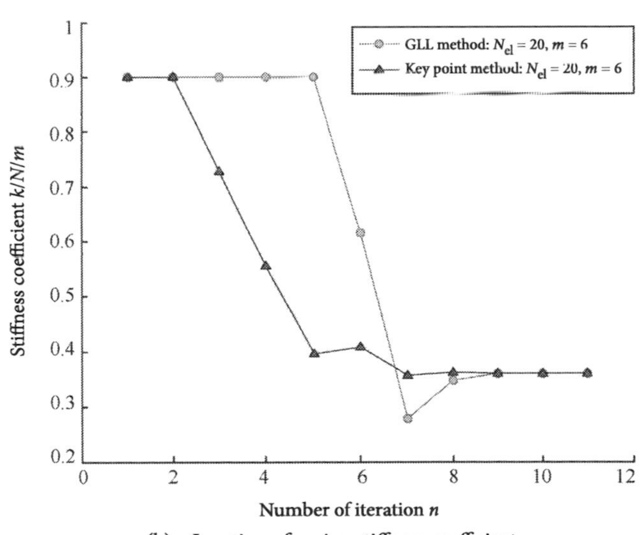

(b) Iteration of spring stiffness coefficient

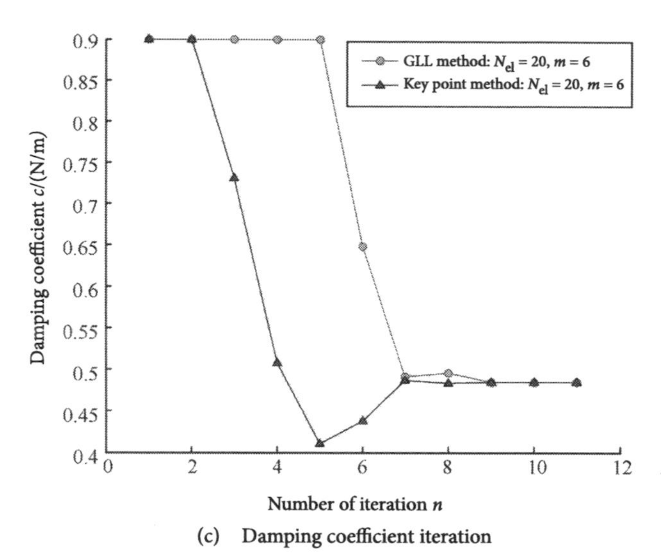

(c) Damping coefficient iteration

Figure 3-5 Comparison of GLL point method and key point method

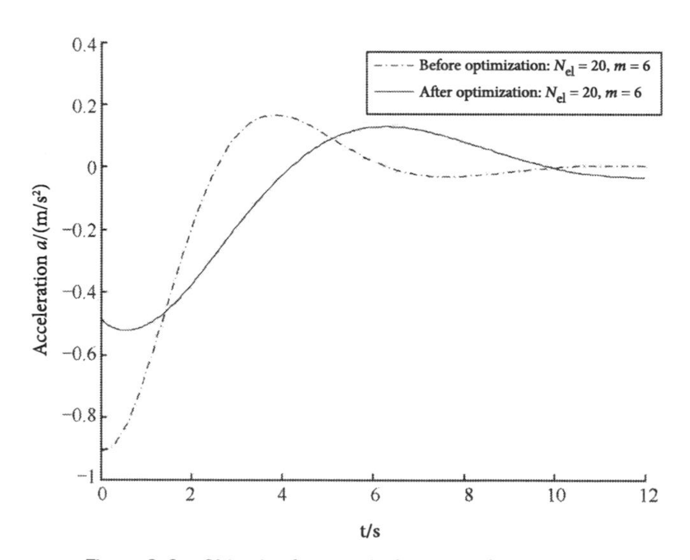

Figure 3-6 Objective function before and after optimization

3.4 Optimal design of linear two-degree-of-freedom shock absorber

In the linear two-degree-of-freedom shock absorber shown in Figure 3-7, $m_1 = 4.534$ kg, $m_2 = 4.534$ kg, $k_1 = 1749.03$ N/m, and the excitation frequency is 1.2 times the natural frequency of the main mass Ω_n, $\omega = 23.57$ rad. To calculate the stiffness coefficient k_2 and damping coefficient c of the damper so that the maximum displacement response of the main mass is minimized, at the same time, it satisfies the vibration space constraint condition, the difference between the displacement response of the shock absorber and the displacement response of the main mass cannot exceed 3 times the maximum displacement response of the main mass, and the steady-state constraint conditions. The model can be expressed by formula (3.5), and the initial conditions of the shock absorber are all zero.

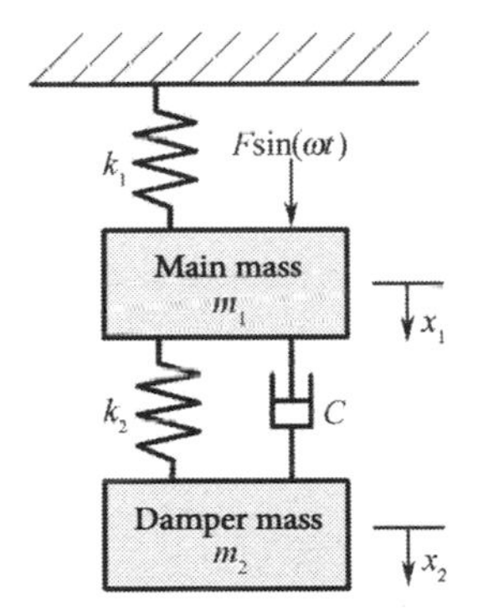

Figure 3-7 Linear two-degree-of-freedom shock absorber

$$\begin{cases} m_1\ddot{x}_1(t) + k_1x_1(t) + k_2[x_1(t) - x_2(t)] + c[\dot{x}_1(t) - \dot{x}_2(t)] = F\sin(\omega t) \\ m_2\ddot{x}_2(t) + k_2[x_2(t) - x_1(t)] + c[\dot{x}_2(t) - \dot{x}_1(t)] = 0 \\ x_1(0) = x_2(0) = 0, \dot{x}_1(0) = \dot{x}_2(0) = 0 \end{cases} \tag{3.5}$$

Introduce the following symbols:

$x_{st} = \sqrt{F/k}$ represents the static displacement of the main mass-produced by the force F; $\Omega_n = \sqrt{k_1/m_1}$ represents the uncoupled natural frequency of the main system; $\omega_n = \sqrt{k_2/m_2}$ represents the uncoupled natural frequency of the damper system; $\bar{\mu} = m_2/m_1$ represents the ratio of the damper mass to the main mass; $f = \omega_n/\Omega_n$ represents the damper and the ratio of the uncoupled natural frequency of the main mass; $\xi = \omega/\Omega_n$ represents the ratio of the excitation frequency to the uncoupled natural frequency of the main mass; $C_c = 2m_2\omega_n$ represents the critical damping; c represents the damping coefficient; $\xi = c/c_c$

represents the damping ratio; $x_1 = x_1(\xi, f, \varsigma)$ represents the displacement of the main mass.

Make $z_i = x_i/x_{st}$ $(i = 1, 2)$, the formula (3.5) is transformed into:

$$\begin{cases} \ddot{z}_1 + \Omega_n^2 z_1 + 2\bar{\mu}\xi\omega_n(\dot{z}_1 - \dot{z}_2) + \bar{\mu}f^2\Omega_n^2(z_1 - z_2) = \Omega_n^2 \sin(\Omega_n \xi t) \\ \ddot{z}_2 + 2\xi\omega_n(\dot{z}_2 - \dot{z}_1) + f^2\Omega_n^2(z_2 - z_1) = 0 \end{cases} \tag{3.6}$$

Make again $x_1 = \dot{z}_1, x_2 = z_1, x_3 = \dot{z}_2, x_4 = z_2$, the formula (3.6) is transformed into:

$$\dot{\boldsymbol{X}} + \boldsymbol{A}_s\boldsymbol{X} = \boldsymbol{F} \tag{3.7}$$

In the formula, $\dot{\boldsymbol{X}} = \left[\dot{x}_1\dot{x}_2\dot{x}_3\dot{x}_4\right]^{\mathrm{T}}, \boldsymbol{X} = \left[\dot{x}_1\dot{x}_2\dot{x}_3\dot{x}_4\right]^{\mathrm{T}}$

$$\boldsymbol{A}_s = \begin{bmatrix} 2\bar{\mu}\xi\omega_n & \Omega_n^2(1+\bar{\mu}f^2) & -2\bar{\mu}\xi\omega_n & -\bar{\mu}f^2\Omega_n^2 \\ -1 & 0 & 0 & 0 \\ -2\xi\omega_n & -f^2\Omega_n^2 & 2\xi\omega_n & f^2\Omega_n^2 \\ 0 & 0 & -1 & 0 \end{bmatrix}$$

$$\boldsymbol{F} = \begin{bmatrix} \Omega_n^2 \sin(\Omega_n \xi t) \\ 0 \\ 0 \\ 0 \end{bmatrix}$$

Then, the shock absorber optimization model is expressed as

$$\begin{cases} \min \max\limits_{t\in[0,T]} \left|x_2(\xi, f, t)\right| \\ \text{s.t.} \ \ \dot{\boldsymbol{X}} + \boldsymbol{A}_s\boldsymbol{X} = \boldsymbol{F} \\ \text{s.t.} \left|x_4(\xi, f, t) - x_2(\xi, f, t)\right| \leqslant 3\max\left|x_2(\xi, f, t)\right| \\ \text{s.t.} \ x_{\text{st4}}(\xi, f, t) \leqslant a \\ \text{s.t.} \left|x_{\text{st4}}(\xi, f, t) - x_{\text{st2}}(\xi, f, t)\right| \leqslant 3a \\ \text{s.t.} \ \xi_{\min} \leqslant \xi \leqslant \xi_{\max} \\ \text{s.t.} \ f_{\min} \leqslant f \leqslant f_{\max} \end{cases} \tag{3.8}$$

According to the literature [118], by introducing the artificial design variable x_5, the formula (3.8) can be transformed into

$$\begin{cases} \overline{\varphi_0}(X) = x \\ \text{s.t.} \quad \dot{X} + A_s X = F \\ \text{s.t.} \left| x_2 \right| \leqslant x_5 \\ \text{s.t.} \left| x_4(\xi, f, t) - x_2(\xi, f, t) \right| \leqslant 3x_5 \\ \text{s.t.} \; x_{st4}(\xi, f, t) \leqslant a \\ \text{s.t.} \left| x_{st4}(\xi, f, t) - x_{st2}(\xi, f, t) \right| \leqslant 3a \\ \text{s.t.} \; \xi_l \leqslant \xi \leqslant \xi_u \\ \text{s.t.} \; f_l \leqslant f \leqslant f_u \end{cases} \tag{3.9}$$

In the formula, a = 0.03048 m; x_{2st} and x_{4st} represent the steady-state displacement response of the main mass and the damper, respectively,
The corresponding expression is

$$\begin{cases} x_{st2} = \sqrt{\dfrac{(2\zeta\xi)^2 + (\zeta^2 - f^2)^2}{x_b}} \\ x_{st4} = \sqrt{\dfrac{(2\zeta\xi)^2 + f^4}{x_b}} \end{cases} \tag{3.10}$$

In the formula, $x_b = (2\zeta\xi)^2(\zeta^2 + \overline{\mu}\zeta^2 - 1)^2 + \left[\overline{\mu}\zeta^2 - (\zeta^2 - 1)(\zeta^2 - f^2)\right]^2$, Damping ratio $\xi \in [10^{-6}, 0.16785]$; $f \in [10^{-4}, 2.0]$; $\overline{\mu} = 0.1$. In the key point method, to find the key point, the golden section method and the parabola method are used for one-dimensional search, and the convergence upper limit of the objective function and the time design variable is set to 1.5×10^{-15}. In order not to miss all possible key points, two one-dimensional searches are carried out in each unit, and all key points found in the unit are required to meet the constraints. Taking one set of data as an example, the iterative process is shown in Figure 3-8 – Figure 3-10. According to literature 65, the known conditions of the gradient projection method are $\overline{\mu} = 0.1$, a = 1.2, $10^{-4} \leq f \leq 2.0$, $10^{-6} \leq \xi \leq 0.16785$, ζ = 1.2; the initial conditions are f = 1.6, ξ = 0.02, d = 0.081, and the optimal result is f = 1.338, ξ = 0.02121, d = 0.0601.

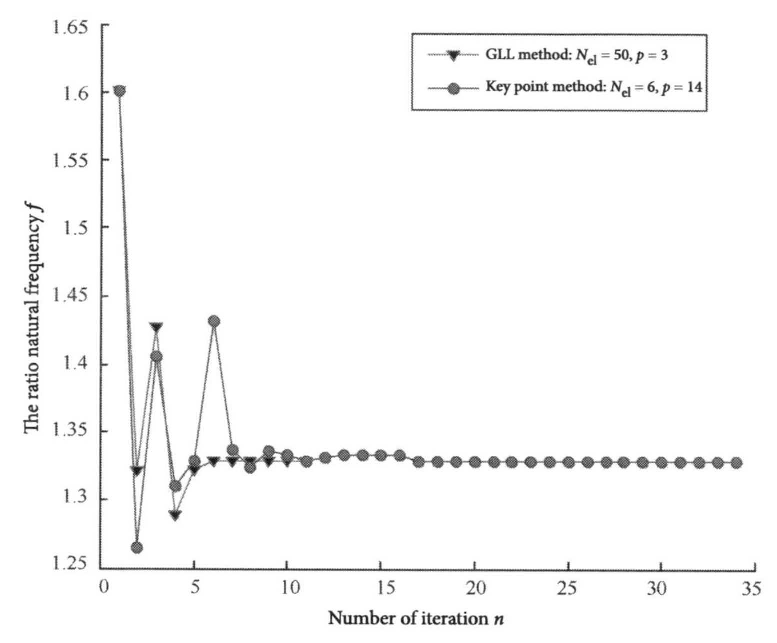

Figure 3-8 Iteration of natural frequency ratio (1)

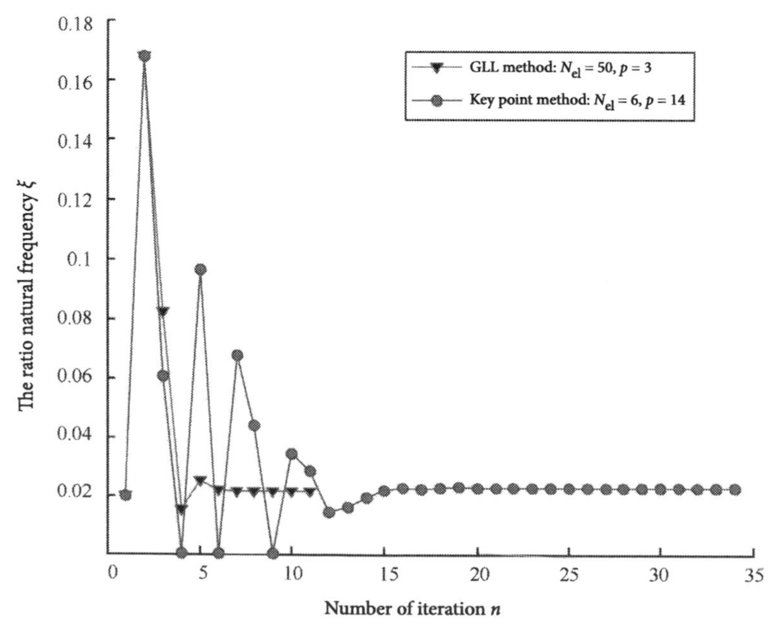

Figure 3-9 Iteration of damping ratio (1)

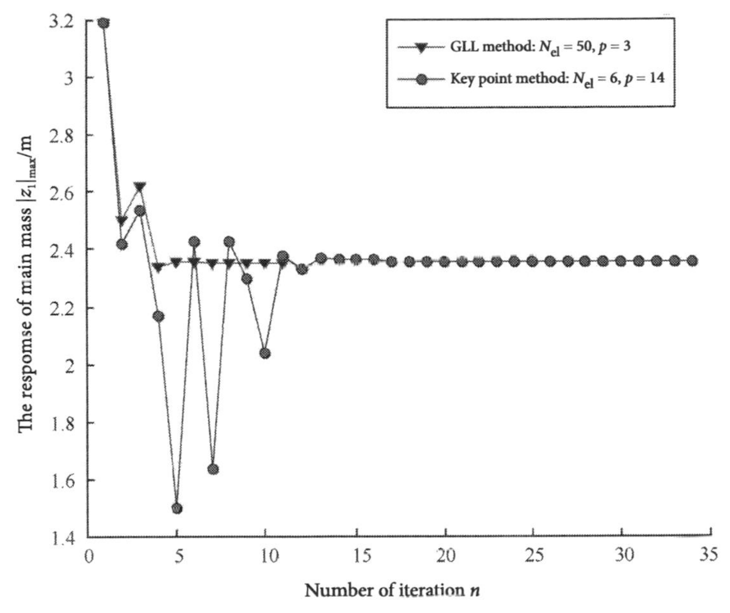

Figure 3-10 Iteration of the main mass dynamic response (1)

The solution results using the GLL point method and key point method are shown in Table 3.5–Table 3.8.

Table 3.5 Best Advantages of GLL Point Method Numerical Test

Number of interpolation points m		Number of units N_{el}					
		10	20	30	40	50	60
1	f	1.4352	1.5072	1.5013	1.2139	1.2139	1.2139
	ξ	0.1679	1×10^{-6}	0.0544	0.1579	0.1579	0.15787
	$\lvert x_1 \rvert$	0.0280	0.0381	0.2263	0.1639	0.1062	0.095949
2	f	1.5072	1.3304	1.3471	1.3275	1.3371	1.3559
	ξ	1×10^{-6}	1×10^{-6}	0.0022	0.0316	1×10^{-6}	1×10^{-6}
	$\lvert x_1 \rvert$	0.3533	0.0570	0.0558	0.0596	0.059658	0.058294
3	f	1.3194	1.3301	1.3287	1.3288	1.3291	1.3563
	ξ	0.0608	0.0098	0.0243	0.0231	0.021252	1×10^{-6}
	$\lvert x_1 \rvert$	0.0632	0.0618	0.0602	0.0597	0.059749	0.058701

(Continued)

Number of interpolation points *m*		Number of units N_{el}					
		10	20	30	40	50	60
4	f	1.3176	1.3507	1.3560	1.3541	1.3563	1.329
	ξ	0.0653	1×10^{-6}	1×10^{-6}	1×10^{-6}	1×10^{-6}	0.022194
	$\lvert x_1 \rvert$	0.0468	0.0578	0.0586	0.0585	0.058667	0.059738
5	f	1.3611	1.3586	1.3286	1.3288	1.3558	1.356
	ξ	1×10^{-6}	0.0009	0.0247	0.0230	1×10^{-6}	1×10^{-6}
	$\lvert x_1 \rvert$	0.0530	0.0581	0.0597	0.0599	0.05867	0.058709
6	f	1.3244	1.3286	1.3573	1.3555	1.3557	1.3563
	ξ	0.0454	0.0247	1×10^{-6}	1×10^{-6}	1×10^{-6}	1×10^{-6}
	$\lvert x_1 \rvert$	0.0583	0.0596	0.0585	0.0586	0.058686	0.058677

Table 3.6 Time-consuming of numerical test of GLL point method (Unit: s)

Number of interpolation points *m*	Number of units N_{el}					
	10	20	30	40	50	60
1	0.5156	0.5313	0.5938	0.9219	0.8125	1.1094
2	0.5	0.7031	6.5781	2.7813	3.625	12.031
3	0.8594	1.4375	3.125	7.9375	10.344	14.469
4	0.7344	3.25	5.3438	9.3125	16.859	32.656
5	0.7188	5.1719	12.6719	16.3281	37.469	34.922
6	1.4844	6.6406	12.0313	21.3906	34.859	42.734

Comparing Table 3.7 with the optimization results, [118] we find that are the best solutions. It can be seen from Table 3.9 that it takes 185.03 s for the GLL point method, and all GLL points are required to meet the constraints. However, it is impossible to find a feasible solution for a small element number solution, and it can only be found if the number of large units and small interpolation points are used. From Table 3.5, it can be found that are the best solutions, which are closer to the optimal data in the literature [118], and they take the only 10.344 s (see Table 3.7). This shows that for the design of linear two-degree-of-freedom shock absorbers, the GLL point method is better than the key point method.

Table 3.7 Best Advantages of Key Point Method Numerical Test

Number of interpolation points m		Number of units N_{el}				
		8	10	12	14	16
1	f	1.2239	1.3304	1.4352	1.3992	2.6235
	ξ	0.1656	1×10^{-6}	0.1679	1×10^{-6}	0.7913
	$\lvert x_1 \rvert$	0.026546	0.044478	0.046129	0.035547	0.396984
2	f	1.2813	1.2270	1.3256	1.3290	1.3801
	ξ	0.1679	0.1679	0.0406	0.0218	0.0088
	$\lvert x_1 \rvert$	0.068892	0.11877	0.074	0.055093	0.057915
3	f	1.2547	1.3301	1.3625	1.3261	1.3289
	ξ	0.1400	0.0095	0.0511	0.0384	0.0389
	$\lvert x_1 \rvert$	0.062794	0.053914	0.055184	0.055062	0.055123
4	f	1.3770	1.3283	1.3287	1.3274	1.3287
	ξ	1×10^{-6}	0.0266	0.0239	0.0319	0.0240
	$\lvert x_1 \rvert$	0.059766	0.055817	0.056708	0.055674	0.056627
5	f	1.3708	1.3567	1.3560	1.3289	1.3288
	ξ	1×10^{-6}	1×10^{-6}	1×10^{-6}	0.0229	0.0229
	$\lvert x_1 \rvert$	0.059492	0.058618	0.058715	0.059883	0.059881
6	f	1.3288	1.3289	1.3289	1.3289	1.3289
	ξ	0.0233	0.0229	0.0229	0.0229	0.0229
	$\lvert x_1 \rvert$	0.059827	0.059878	0.059883	0.059883	0.059883

Table 3.8 Numerical test time consumption of key point method

Number of interpolation points m	Number of units N_{el}				
	8	10	12	14	16
1	31.95	7.20	5.36	32.05	5.42
2	70.44	8.41	38.16	63.95	55.58
3	103.17	56.23	114.58	117.45	87.10
4	25.27	108.36	135.66	166.70	109.50
5	157.14	180.91	159.78	202.27	114.60
6	84.08	115.17	165.92	185.03	230.50

In fact, from the time-consuming analysis, it is also possible to provide an analytical gradient function or a simple numerical gradient function for the SQP algorithm, so that the number of evaluations of the function can be significantly reduced. [119] The SQP algorithm uses the difference method to calculate the gradient, and each iteration requires multiple simulation function evaluations (the number of evaluations is equal to twice the dimensionality of the design variable multiplied by the number of instances, and then added a 1), and the optimization solution is very time-consuming. [120] For dynamic response optimization based on the spectral element method, every time the function value is evaluated and the spectral element method must be used to update it. Obviously, the more the number of units and the number of interpolation points, the more function evaluation times, and the more time it takes; reducing the number of function evaluation times will naturally reduce the time consumption. For the linear two-degree-of-freedom shock absorber studied in this book, we take $m = 3$, $N_{el} = 50$ and $m = 3$, $N_{el} = 60$ from Table 3.5 to optimize the initial value to study whether the GLL point is at the maximum point. The results are shown in Figure 3-11 and Figure 3-12. It can be seen from Figure 3-11 and Figure 3-12 that the closer it is to the maximum point, the higher the accuracy of convergence; the farther away from the maximum point, the lower the accuracy of convergence. Table 3.9 compares the data in the reference with the data obtained in this book, from which we can see the correctness and feasibility of the method in this book.

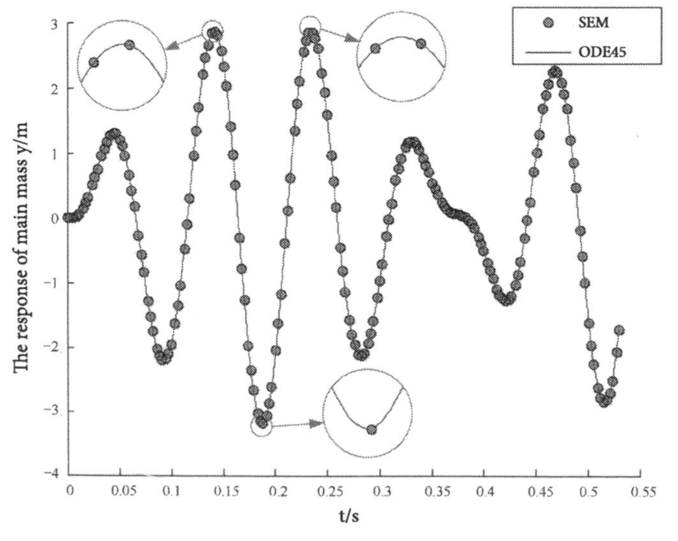

Figure 3-11 Research of the GLL point at the maximum point (the number of units is 50, and the number of interpolation points is 3)

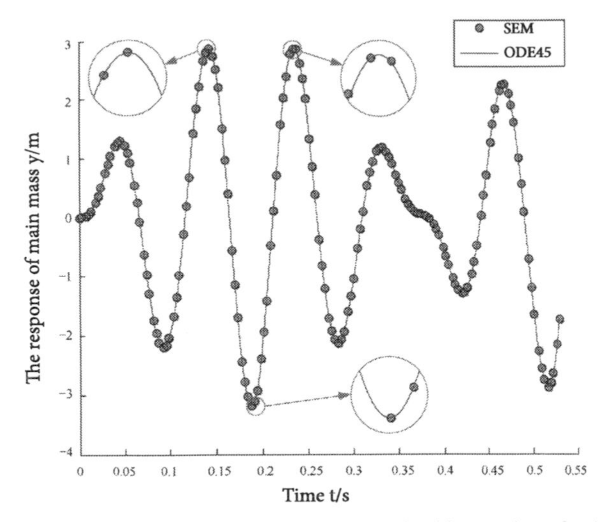

Figure 3-12 Research of the GLL point at the maximum point (the number of units is 60, the number of interpolation points is 3)

NOTE: ODE45 is a method in Matlab.

Table 3.9 Comparison of reference data and data obtained in this book

Parameter	Book data	Literature optimization result data				
		Literature [118]	Literature [121]	Literature [122]	Literature [123]	Literature [113]
Natural frequency ratio f	1.3289	1.338	1.3277	1.3277	1.3312	1.3597
Damping coefficient ratio ξ	0.0229	0.02121	0.03054	0.03058	0.02758	0.0184
Large displacement response $\lvert x_1 \rvert$	0.059883	0.060122	0.059858	0.059847	0.059842	0.060335

From Table 3.10 to Table 3.13, it can be seen that the GLL point method has great advantages in terms of speed, but in terms of accuracy, it is less superior. In Table 3.10, the parameters on the lower right have found the accuracy of the most advantages, and the damping ratios corresponding to the parameters on the upper left found are also very small, which is unreasonable for design. In the GLL point method test results, the number of units is 100, 200, 300, 400, and the number of interpolation points is 3. At the same time, the number of units is 100, 200, and the number of interpolation points is

6, while the number of units is 100, and the number of interpolation points is 10. These test results are unreasonable and the rest of the results are reasonable, as shown in Table 3.11. In the key point method test results, all the results are reasonable, as shown in Table 3.12. In terms of time-frame, the GLL point method's least time-consuming best point is 21.359375 s, while the corresponding number of units is 300, and the number of interpolation points is 6. The most time-consuming is 370.375000 s, the corresponding number of units is 600, and the number of interpolation points is 10, as shown in Table 3.11. For the key point method, the minimum time-consuming point is 256.32812 s, the corresponding number of units is 100, and the number of interpolation points is 6. At the same time, the longest time-consuming point is 2171.03125 s, the number of corresponding units is 800, and the number of interpolation points is 6, as shown in Table 3.13. It is shown from Figure 3-13 to Figure 3-15 that the GLL point method is better than the key point method, and it can quickly and stably find the best point. But in terms of accuracy, the key point method is better than the GLL point method.

Table 3.10 Best advantages of GLL point method numerical test

Number of interpolation points m		Number of units N_{el}					
		100	200	300	400	600	800
3	f	1.356068	1.356083	1.356037	1.356046	1.328857	1.328857
	ξ	0.000001	0.000001	0.000001	0.000001	0.022917	0.022918
	$\lvert x_1 \rvert$	0.058681	0.058711	0.058712	0.058709	0.059881	0.059881
6	f	1.355815	1.356080	1.328860	1.328858	1.328858	1.328857
	ξ	0.000001	0.000001	0.022894	0.022911	0.022908	0.022912
	$\lvert x_1 \rvert$	0.058702	0.058712	0.059880	0.059882	0.059882	0.059882
10	f	1.356054	1.328860	1.328858	1.328857	1.328857	1.328507
	ξ	0.000001	0.022893	0.022906	0.022914	0.022912	0.025423
	$\lvert x_1 \rvert$	0.058713	0.059880	0.059881	0.059882	0.059882	0.060091

Table 3.11 Time consumption and number CPU iterations CPU the point method

Number of interpolation points m		Number of units N_{el}					
		100	200	300	400	600	800
3	t	5.765625	4.125000	5.109375	6.171875	31.750000	54.265625
	n	9	12	10	10	38	52
6	t	4.578125	68.937500	21.359375	49.890625	24.125000	36.796875
	n	11	101	19	38	11	12
10	t	8.437500	29.828125	22.718750	33.437500	370.375000	610.937500
	n	11	22	11	12	106	124

NOTE: The unit of CPU time is s.

Table 3.12 The most advantages of the key point method numerical test

Number of interpolation points m		Number of units N_{el}					
		100	200	300	400	600	800
3	f	1.328858	1.328857	1.328857	1.328857	1.328857	1.328857
	ξ	0.022909	0.022912	0.022912	0.022912	0.022912	0.022912
	$\|x_1\|$	0.059884	0.059882	0.059882	0.059882	0.059882	0.059882
6	f	1.328857	1.328857	1.328857	1.328857	1.328857	1.328857
	ξ	0.022912	0.022912	0.022912	0.022912	0.022912	0.022912
	$\|x_1\|$	0.059882	0.059882	0.059882	0.059882	0.059882	0.059882
10	f	1.328857	1.328857	1.328857	1.328857	1.328857	1.328857
	ξ	0.022912	0.022912	0.022912	0.022912	0.022912	0.022912
	$\|x_1\|$	0.059882	0.059882	0.059882	0.059882	0.059882	0.059882

Table 3.13 Critical point method CPU time consumption and number of iterations

Number of interpolation points m		Number of units N_{el}					
		100	200	300	400	600	800
3	t	349.90625	558.20312	807.078125	1263.6562	1000.71875	1163.71875
	n	32	37	35	37	25	24

(Continued)

Number of interpolation points m		Number of units N_{el}					
		100	200	300	400	600	800
6	t	256.32812	444.17187	591.406250	754.43750	1140.82812	2171.03125
	n	26	30	28	28	27	31
10	t	404.4062	567.3750	656.68750	782.125000	1199.00000	2061.46875
	n	37	31	26	26	27	27

Note: The unit of CPU time is s.

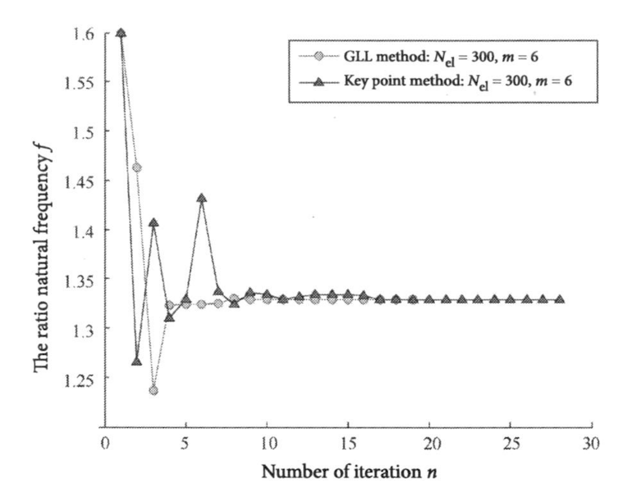

Figure 3-13 Iteration of natural frequency ratio (2)

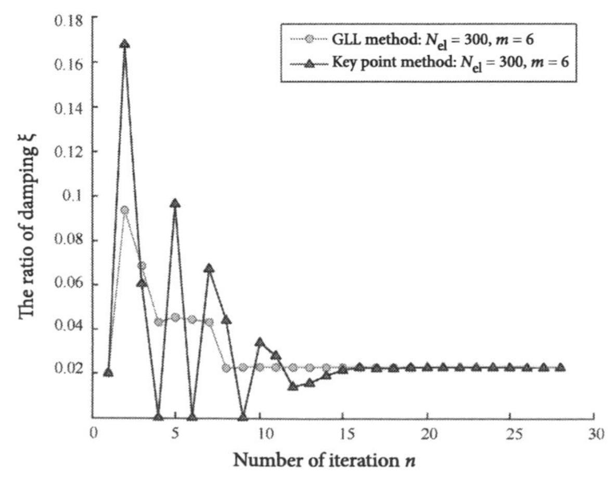

Figure 3-14 Iteration of damping ratio (2)

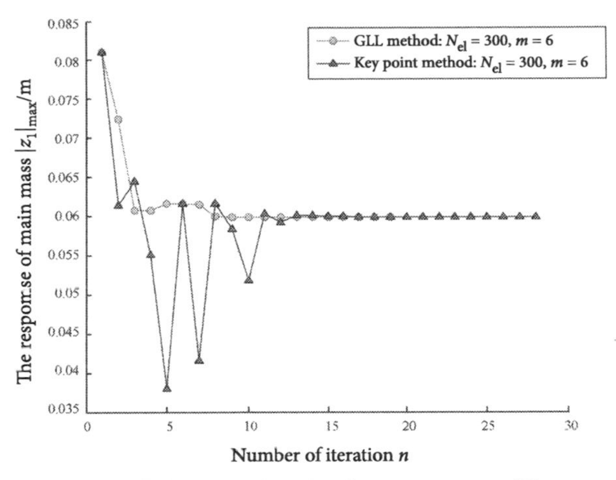

Figure 3-15 Iteration of main mass response (2)

3.5 Optimal design of dynamic response of automobile suspension system

The automobile system can be approximately simplified to a five-degree-of-freedom dynamic model as shown in Figure 3-16. In Figure 3-16, m_1 is the mass of the driver and his seat. It is supported on the car body by a spring with a stiffness coefficient of k_1 and a damper with a damping coefficient of c_1. Meanwhile, the masses of the car body, front and rear axles, and wheels are m_2, m_4, and m_5 respectively. The car body is supported by springs with stiffness coefficients k_2 and k_3 and dampers with damping coefficients c_2 and c_3 connected to the front and rear axles. The k_4, k_5 and c_4, c_5 respectively represent the stiffness coefficient and damping coefficient of the tire. The moment of inertia of the car body to its center of mass is represented by I, while the L represents the front and rear wheel distance. The points $f_1(t)$ and $f_2(t)$ represent the displacement functions of the front and rear wheels of the car caused by the unevenness of the road surface. The point $y_i(t)$ is the generalized coordinate of five degrees of freedom.

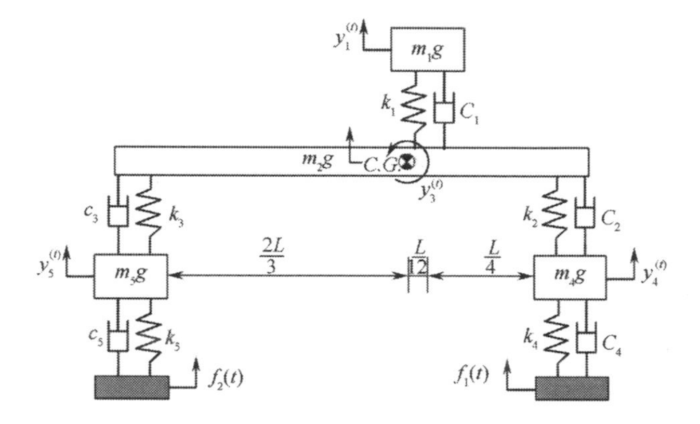

Figure 3-16 Simplified dynamic model of automobile master degrees of freedom

This optimization problem only takes the spring stiffness coefficient k_i and damping coefficient c_i (i = 1, 2, 3) related to the suspension system in the model as the design variables and takes the minimum transient dynamic response of the driver and passenger seat as the goal. The system state equation and amplitude are constraints, as described in the following.

1. Choosing design variables

This optimization problem only takes the spring stiffness coefficient k_i and damping coefficient c_i (i = 1, 2, 3) related to the suspension system in the model as design variables, namely $\boldsymbol{X} = (k_1,k_2,k_3,c_1,c_2,c_3)^{\mathrm{T}}$.

2. Establishing the objective function

To minimize the transient dynamic response of the driver and passenger seat, the dynamic response of the mechanical system is optimized. That is, the structural design of the car suspension system requires the car to operate at various speeds and road conditions. And the maximum acceleration response of the driver and passenger seat is the smallest, which can be expressed as:

$$f(X,t) = \min[\max|\ddot{y}_1(t)|]$$

3. Satisfying the given constraints

(1) Meeting the constraints of the system state equation

The differential equation of system motion is established by the Lagrange method, and then transformed into the state equation of the system, as shown in equation (3.11).

$$M\ddot{y} + C\dot{y} + Ky = bF(t) \tag{3.11}$$

In the formula:

$$M = \begin{bmatrix} m_1 & 0 & 0 & 0 & 0 \\ 0 & m_2 & 0 & 0 & 0 \\ 0 & 0 & I & 0 & 0 \\ 0 & 0 & 0 & m_4 & 0 \\ 0 & 0 & 0 & 0 & m_5 \end{bmatrix}$$

$$b = \begin{bmatrix} 0 & 0 & 0 & 0 & 0 \\ 0 & 0 & 0 & 0 & 0 \\ 0 & 0 & 0 & 0 & 0 \\ 0 & k_4 & c_4 & 0 & 0 \\ 0 & 0 & 0 & k_5 & c_5 \end{bmatrix}$$

$$k = \begin{bmatrix} k_1 & -k_1 & -\frac{Lk_1}{12} & 0 & 0 \\ -k_1 & k_1 + k_2 + k_3 & \frac{L}{3}\left(\frac{1}{4}k_1 + k_2 - 2k_3\right) & -k_2 & -k_3 \\ -\frac{Lk_1}{12} & \frac{L}{3}\left(\frac{1}{4}k_1 + k_2 - 2k_3\right) & \frac{L2}{9}\left(\frac{1}{16}k_1 + k_2 + 4k_3\right) & -\frac{Lk_2}{3} & \frac{2Lk_3}{3} \\ 0 & -k_2 & -\frac{Lk_3}{3} & k_2 + k_4 & 0 \\ c_1 & -k_3 & \frac{Lk_3}{3} & 0 & k_3 + k_4 \end{bmatrix}$$

$$C = \begin{bmatrix} c_1 & -c_1 & -\frac{Lc_1}{12} & 0 & 0 \\ -c_1 & c_1 + c_2 + c_3 & \frac{L}{3}\left(\frac{1}{4}c_1 + c_2 - 2c_3\right) & -c_2 & -c_3 \\ -\frac{Lc_1}{12} & \frac{L}{3}\left(\frac{1}{4}c_1 + c_2 - 2c_3\right) & \frac{L2}{9}\left(\frac{1}{16}c_1 + c_2 + 4c_3\right) & -\frac{Lc_2}{3} & \frac{2Lc_3}{3} \\ 0 & -c_2 & -\frac{Lc_3}{3} & c_2 + c_4 & 0 \\ c_1 & -c_3 & \frac{2Lc_3}{3} & 0 & c_3 + c_4 \end{bmatrix}$$

$F(t) = \left[0 \;\; f_1(t) \;\; \dot{f}_1(t) \;\; f_2(t) \;\; \dot{f}_2(t)\right]^{\mathrm{T}}$, $y = \left[y_1 \;\; y_2 \;\; y_3 \;\; y_4 \;\; y_5\right]^{\mathrm{T}}$, $\dot{y} = \left[\dot{y}_1 \;\; \dot{y}_2 \;\; \dot{y}_3 \;\; \dot{y}_4 \;\; \dot{y}_5\right]^{\mathrm{T}}$, $\ddot{y} = \left[\ddot{y}_1 \;\; \ddot{y}_2 \;\; \ddot{y}_3 \;\; \ddot{y}_4 \;\; \ddot{y}_5\right]^{\mathrm{T}}$.

Equation (3.11) can be simplified to

$$\ddot{y} + C\dot{y} + Ky = bF(t) \tag{3.12}$$

In the formula

$$b = \begin{bmatrix} 0 & 0 & 0 & 0 & 0 \\ 0 & 0 & 0 & 0 & 0 \\ 0 & 0 & 0 & 0 & 0 \\ 0 & \frac{k_4}{m_4} & \frac{c_4}{m_4} & 0 & 0 \\ 0 & 0 & 0 & \frac{k_5}{m_5} & \frac{c_5}{m_5} \end{bmatrix}$$

$$K = \begin{bmatrix} \frac{k_1}{m_1} & \frac{-k_1}{m_1} & -\frac{Lk_1}{12m_1} & 0 & 0 \\ \frac{-k_1}{m_2} & \frac{k_1+k_2+k_3}{m_2} & \frac{L}{3m_2}\left(\frac{1}{4}k_1+k_2-2k_3\right) & \frac{-k_2}{m_2} & \frac{-k_3}{m_2} \\ -\frac{Lk_1}{12m_3} & \frac{L}{3m_3}\left(\frac{1}{4}k_1+k_2-2k_3\right) & \frac{L2}{9m_3}\left(\frac{1}{16}k_1+k_2+4k_3\right) & -\frac{Lk_2}{3m_3} & \frac{2Lk_3}{3m_3} \\ 0 & \frac{-k_2}{m_3} & -\frac{Lk_3}{3m_3} & \frac{k_2+k_4}{m_3} & 0 \\ c_1 & \frac{-k_3}{m_3} & \frac{Lk_3}{3m_3} & 0 & \frac{k_3+k_4}{m_3} \end{bmatrix}$$

$$C = \begin{bmatrix} \frac{c_1}{m_1} & \frac{-c_1}{m_1} & -\frac{Lc_1}{12m_1} & 0 & 0 \\ \frac{-c_1}{m_2} & \frac{c_1+c_2+c_3}{m_2} & \frac{L}{3m_2}\left(\frac{1}{4}c_1+c_2-2c_3\right) & \frac{-c_2}{m_2} & \frac{-c_3}{m_2} \\ -\frac{Lc_1}{12m_3} & \frac{L}{3m_3}\left(\frac{1}{4}c_1+c_2-2c_3\right) & \frac{L2}{9m_3}\left(\frac{1}{16}c_1+c_2+4c_3\right) & -\frac{Lc_2}{3m_3} & \frac{2Lc_3}{3m_3} \\ 0 & \frac{-c_2}{m_4} & -\frac{Lc_2}{3m_4} & \frac{c_2+c_4}{m_4} & 0 \\ 0 & \frac{-c_3}{m_5} & \frac{2Lc_3}{3m_5} & 0 & \frac{c_3+c_4}{m_5} \end{bmatrix}$$

Make $x = y$, $z = \dot{y}$, Then the formula (3.12) can be transformed into a first-order differential equation system:

$$\begin{cases} \dot{\boldsymbol{x}} - z = 0 \\ \dot{z} + \boldsymbol{C}_z + \boldsymbol{Kx} = \boldsymbol{bF}(t) \end{cases} \tag{3.13}$$

Written in matrix form as:

$$\begin{bmatrix} \dot{x} \\ \dot{z} \end{bmatrix} + \begin{bmatrix} 0 & -\boldsymbol{I}_{5\times5} \\ \boldsymbol{K} & \boldsymbol{C} \end{bmatrix} \begin{bmatrix} x \\ z \end{bmatrix} = \begin{bmatrix} 0 \\ \boldsymbol{bF}(t) \end{bmatrix} \tag{3.14}$$

(2) Satisfying functional constraints

To improve the comfort of the driver's (passenger) seat, the maximum absolute acceleration is required to be minimum,

$$|\ddot{y}_1(t)| \leqslant d \tag{3.15}$$

In the formula, d is an artificial variable. At the same time, it is required to limit its amplitude to a certain range, that is, the relative displacement between the car body and the driver's seat is restricted as

$$\left| y_2(t) + \frac{L}{12} y_3(t) - y_1(t) \right| \leqslant \theta_1 \tag{3.16}$$

The relative displacement between the car body and the front wheel is constrained as

$$\left| y_4(t) - y_2(t) - \frac{L}{3} y_3(t) \right| \leqslant \theta_2 \tag{3.17}$$

The relative displacement between the car body and the rear wheel is constrained as

$$\left| y_5(t) - y_2(t) - \frac{2L}{3} y_3(t) \right| \leqslant \theta_3 \tag{3.18}$$

The relative displacement between the road surface and the front wheel is constrained as

$$|y_4(t) - f_1(t)| \leqslant \theta_4 \tag{3.19}$$

The relative displacement between the road surface and the rear wheel is constrained as

$$|y_5(t) - f_2(t)| \leqslant \theta_5 \tag{3.20}$$

In the formula, $\theta_1 - \theta_5$ are the maximum allowable relative displacement.

(3) Satisfying the constraints of the range of design variables

$$X = \left[k_1, k_2, k_3, c_1, c_2, c_3, d\right]^{\mathrm{T}}$$

4. *Establishing dynamic response optimization mathematical model*

$$\begin{cases} \overline{\varphi}_0 = d \\ \left|\ddot{y}_1(t)\right| \leqslant d \\ \begin{bmatrix} \dot{x} \\ \dot{z} \end{bmatrix} + \begin{bmatrix} 0 & -\boldsymbol{I}_{5\times5} \\ \boldsymbol{K} & \boldsymbol{C} \end{bmatrix} \begin{bmatrix} \boldsymbol{x} \\ z \end{bmatrix} = \begin{bmatrix} 0 \\ \boldsymbol{b}\boldsymbol{F}(t) \end{bmatrix} \\ \left| y_2(t) + \dfrac{L}{12} y_3(t) - y_1(t) \right| \leqslant \theta_1 \\ \left| y_4(t) - y_2(t) - \dfrac{L}{3} y_3(t) \right| \leqslant \theta_2 \\ y_5(t) - y_2(t) - \dfrac{2L}{3} y_3(t) \leqslant \theta_3 \\ \left|y_4(t) - f_1(t)\right| \leqslant \theta_4,\ \left|y_5(t) - f_2(t)\right| \leqslant \theta_5 \\ x_{\min} \leqslant x \leqslant x_{\max}, z_{\min} \leqslant z \leqslant z_{\max},\ d \geqslant 0 \end{cases} \tag{3.21}$$

3.5.1 Road condition one

The vehicle speed is v = 11.43 m/s, the road surface excitation frequency is $\omega_i = \dfrac{\pi v}{L_i}$,

L_1 = 9.144 m, L_2 = 3.6576 m the unknown time interval for the rear wheels to reach the front wheels is t_σ = 0.2667 s, and the vertical displacement function of the front and rear wheels aroused by the unevenness of the road is $f_1(t)$, $f_2(t)$, and their values are

$$f_1(t) = \begin{cases} y_0\left[1 - \cos\omega_i(t - t^{i-1})\right], t^{i-1} \leqslant t \leqslant t^i, i = 1, 3, \cdots, 2n-1, 0 \leqslant t \leqslant t_1 \\ y_0\left[1 + \cos\omega_i(t - t^{i-1})\right], t^{i-1} \leqslant t \leqslant t^i, i = 1, 3, \cdots, 2n, 0 \leqslant t \leqslant t_1 \end{cases} \tag{3.22}$$

$$f_2(t) = f_1(t - t_\sigma), t_\sigma \leqslant t \leqslant t_1 + t_\sigma \tag{3.23}$$

In the formula, ω_1 = 1.25π rad/s, ω_2 = 3.125π rad/s.

The road surface profile of road condition 1 is shown in Figure 3-17.

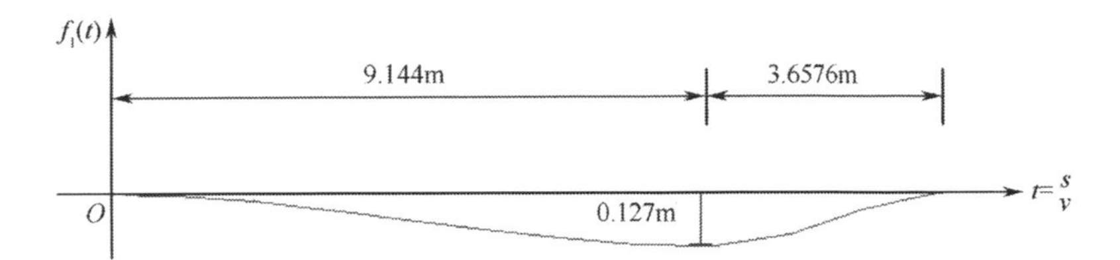

Figure 3-17 Road profile of road condition 1

Substitute the first type of road condition functions $f_1(t)$, and $f_2(t)$, namely formula (3.22) and formula (3.23) into formula (3.21) and apply the spectral element method to solve them. When optimizing linear single-degree-of-freedom and two-degree-of-freedom systems in this chapter, it is concluded that when dealing with multi-degree-of-freedom dynamic problems related to time constraints, the GLL point method has advantages in terms of its time-consumption, and the key point method has advantages in accuracy. Therefore, the GLL point method and the key point method are respectively used here, and the results are shown in Figure 3-18 – Figure 3-20 and Table 3.14 – Table 3.18.

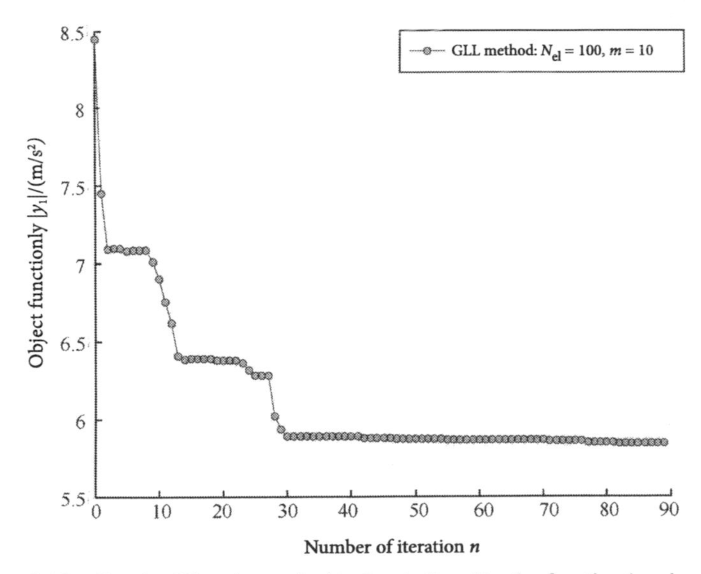

Figure 3-18 Use the GLL point method to iterate the objective function (road condition 1)

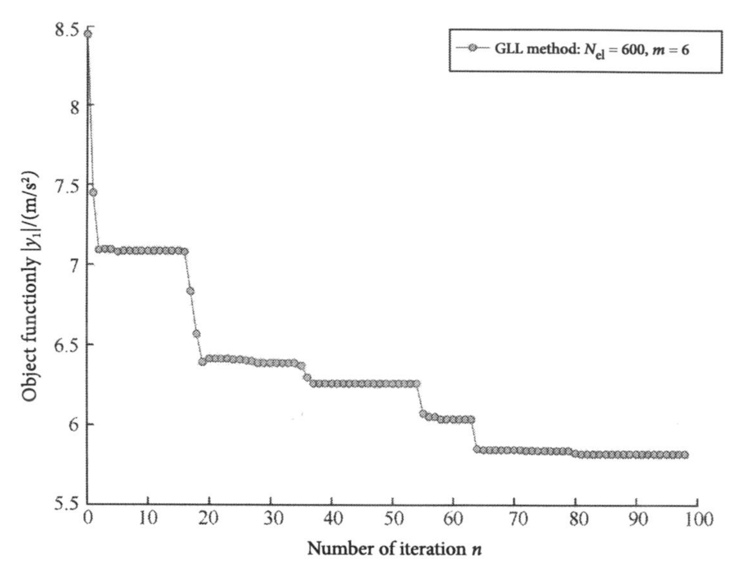

Figure 3-19 Using the key point method to iterate the objective function (road condition 1)

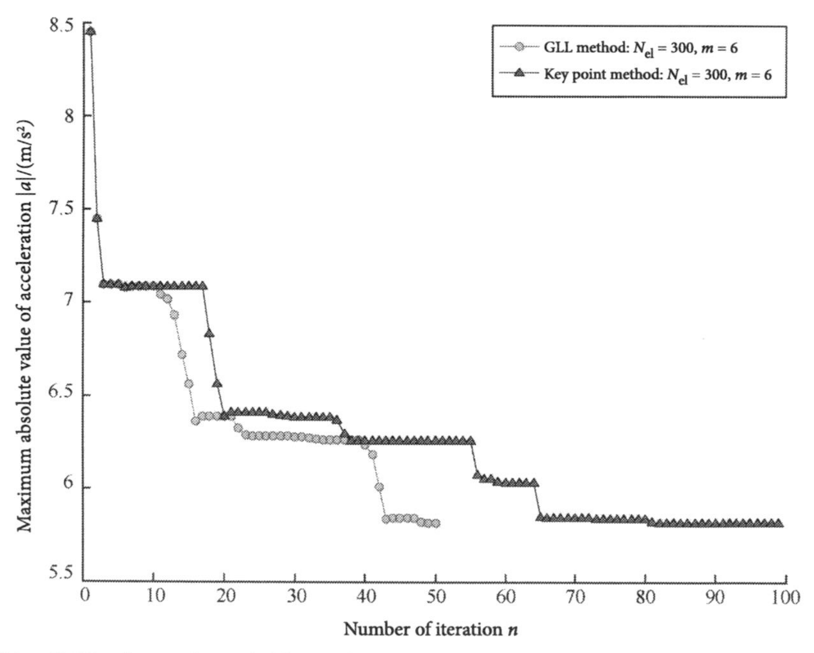

Figure 3-20 Comparison of objective function iteration process under the two methods (road condition 1)

Table 3.14 Advantages of GLL Point Method Numerical Test (Road Condition 1)

Number of interpolation points m		Number of units N_{el}								
		10	20	60	100	200	300	400	600	800
3	k_1	8756.3	8756.3	8756.3	11512.577	1313.764	11368.397	10621.090976	11352.775167	11304.571321
	k_2	35025.2	35025.2	35025.2	35025.2	35025.2	35025.2	35025.200000	35025.200000	35025.200000
	k_3	56096.965	52896.294	54380.17	54236.69	54012.453	53880.53	54296.810450	54092.35177	5 4015.669859
	c_1	5874.6508	5335.082	7439.2012	5311.5278	8313.5847	8672.8911	8749.225365	8355.953880	8594.349953
	c_2	8698.1375	8683.9768	8744.6034	8732.3245	8751.1485	8751.0193	8758.980864	8753.773509	8753.261642
	c_3	875.63	875.63	875.63	875.63	875.63	875.63	875.630000	875.630000	875.630000
	d	5.635387	5.679881	5.820114	5.861439	5.842738	5.842372	5.836000	5.843105	5.842101
6	k_1	16829.493	8756.3	11375.591	11147.023	11313.574	11190.163	11319.587419	8756.314798	10852.478996
	k_2	52441.299	35025.2	35025.2	35025.2	35025.2	35025.2	35025.200000	35025.200000	35025.200000
	k_3	52444.881	52325.981	54058.037	53953.222	53887.454	54057.191	54167.781032	54303.52419	1 4105.030017
	c_1	8369.2674	8756.3	8595.8982	8460.695	8690.769	8739.3201	8756.300000	8724.263448	8756.300000
	c_2	6921.9346	8711.5346	8752.9325	8750.414	8751.0811	8754.734	8757.287613	8756.446953	8755.382237
	c_3	3742.2472	875.63	875.63	875.63	875.63	875.63	875.630000	875.630000	875.630000
	d	6.391697	5.783691	5.840103	5.840892	5.841976	5.840749	5.841821	5.820047	5.837981

(Continued)

Number of interpolation points m		Number of units N_{el}								
		10	20	60	100	200	300	400	600	800
10	k_1	10463.497	10988.386	11465.389	11354.467	10781.71	11333.207	11420.985152	16119.914150	11346.898801
	k_2	35025.2	35025.2	35025.2	35025.2	35025.2	35025.2	35025.200000	50497.41577	2 5025.200000
	k_3	53007.71	54250.642	54200.764	53769.299	54068.151	54080.893	53796.788090	52197.68288	7 3926.860607
	c_1	6989.117	7134.3363	5752.0915	8756.2999	8756.2999	8682.4604	8756.300000	8756.300000	8756.300000
	c_2	8717.1598	8730.6553	8738.76	8748.2939	8754.5056	8755.1213	8749.721132	8822.499714	8752.349274
	c_3	875.63	875.63	875.63	875.63	875.63	875.63	875.630000	875.630000	875.630000
	d	5.784874	5.826069	5.858013	5.842045	5.837389	5.842109	5.842758	6.285569	5.842124

NOTE: The unit of k_i ($i = 1, 2, 3$) in the table is N/m; the unit of c_i ($i = 1, 2, 3$) is N/(m/s); the unit of d is m/s^2. Same as below.

Table 3.15 GLL point method numerical test CPU time consumption and number of iterations (road condition 1)

Number of interpolation points m		Number of units N_{el}								
		10	20	60	100	200	300	400	600	800
3	t	21.3125	32.9375	55.34375	45.5625	111.39063	105.03125	235.062500	286.750000	305.218750
	n	100	134	94	48	63	46	77	68	56
6	t	7.875	33.59375	59.859375	158.04688	253.07813	226.28125	462.109375	569.921875	1251.31250
	n	30	73	50	81	72	48	64	50	70
10	t	30.03125	58.65625	155.89063	374.625	737.3125	692.375	1217.343750	1217.421875	2657.67187
	n	64	69	60	90	96	55	71	46	72

Table 3.16 Advantages of Key Point Method Numerical Test (Road Condition 1)

Number of interpolation points m		Number of units N_{el}								
		10	20	60	100	200	300	400	600	800
3	k_1	8756.3000	8756.3000	8756.3000	8756.3000	8756.3000	8756.3000	8756.300000	8756.300000	8756.300000
	k_2	35025.200	35025.200	35025.200	35025.200	35025.200	35025.200	35025.200000	35025.200000	35025.200000
	k_3	50346.570	35025.200	35025.200	50358.411	52366.735	52807.800	52798.648129	53765.972970	54195.095496
	c_1	8756.300	1676.1812	3645.4055	8682.0458	8266.9709	8173.8100	6342.630820	8134.424942	7127.824714
	c_2	13133.195	8216.2125	7735.2723	8613.9762	8680.5405	8712.3996	8702.972389	8740.327862	8743.330250
	c_3	875.63000	9454.7825	1257.1622	875.63000	875.63000	875.63000	875.630000	875.630000	875.630000
	d	1.963638	3.910859	5.478822	5.698961	5.787683	5.805585	5.807309	5.815653	5.814718
6	k_1	8756.3000	26698.367	8756.3000	8756.3000	8756.3000	8756.3000	8756.300000	8756.300000	8756.300000
	k_2	35025.200	35025.200	35025.200	35025.200	35025.200	35025.200	35025.200000	35025.200000	35025.200000
	k_3	35025.200	35025.200	39705.641	35025.200	52388.835	52907.790	52879.662203	53689.828180	54186.287798
	c_1	8756.3000	1000.4292	8756.3000	3943.0067	8268.6120	8174.8995	8145.481357	8133.434345	8132.720909
	c_2	14010.080	7602.9390	8093.1504	8081.0408	8681.0326	8714.5676	8717.845383	8738.745807	8750.008370
	c_3	4202.5142	9525.4147	875.63000	997.60482	875.63000	875.63000	875.630000	875.630000	875.630000
	d	1.925773	4.018860	5.593787	5.669711	5.787600	5.805577	5.811567	5.815670	5.817054

(Continued)

Number of interpolation points m		Number of units N_{el}								
		10	20	60	100	200	300	400	600	800
10	k_1	8756.3000	28163.175	8756.3000	8756.3000	8756.3000	8756.3000	8756.300000	8756.300000	8756.300000
	k_2	35025.200	35025.200	35025.200	35025.200	35025.200	35025.200	35025.200000	35025.200000	35025.200000
	k_3	35025.200	35025.200	35025.200	50571.908	52409.528	53301.292	52827.155084	64340.450777	54218.661737
	c_1	8756.3000	930.22129	3631.4489	8675.3064	8268.8297	8179.3519	8144.903038	8401.268680	8133.165384
	c_2	14010.080	7394.7033	7744.7271	8619.1501	8681.4695	8723.0446	8716.731020	8928.444706	8750.678298
	c_3	4281.7080	6917.9697	1242.5194	875.63000	875.63000	875.63000	875.630000	875.630000	875.630000
	d	1.956334	4.006835	5.484301	5.698983	5.787598	5.805534	5.811576	5.811778	5.817046

Table 3.17 Critical point method numerical test CPU time consumption and iteration number (road condition 1)

Number of interpolation points m		Number of units N_{el}								
		10	20	60	100	200	300	400	600	800
3	t	275.890	547.625	3702.125	1553.796	3181.859	6046.281	5639.187500	12152.203125	10085.734375
	n	104	142	404	110	121	161	116	166	108
6	t	463.031	959.312	4899.640	1687.250	4034.890	6417.500	9405.093750	7978.203125	10823.031250
	n	195	233	500	111	139	154	171	99	98
10	t	475.406	789.125	2184.203	1512.609	3494.187	4649.609	11047.390625	49741.078125	39162.531250
	n	167	143	186	82	101	92	168	500	293

Table 3.18 Comparison of book results and literature results

Parameter	Results of this book			Literature [65] Results	
	Initial value	Optimal value		Initial value	Optimal value
		GLL point method	Key point method		
k_1	17512.6	11375.60	8756.30	17512.6	8756.30
k_2	52537.8	35025.20	35025.20	52537.8	35025.25
k_3	52537.8	54058.03	52879.66	52537.8	42362.98
c_1	1751.26	8595.90	8145.48	1751.26	2257.37
c_2	4378.15	8752.93	8717.84	4378.15	13575.77
c_3	4378.15	875.63	875.63	4378.15	14010.08
D (Objective function)	8.448	5.840	5.811567	8.448	6.538

It can be seen from Table 3.14 that when the number of units is $N_{el} = 10$ and the number of interpolation points is $m = 3$, a better result is obtained, the 70 iterations take only 89 s. When the number of units is $N_{el} = 20$ and the number of interpolation points is $m = 6$, it takes 501 s to iterate 200 times, and the target value is 5.765 m/s^2. Therefore, more units and interpolation points do not equal a better result.

It can be seen from Table 3.16 and Table 3.17, except for the cases where the number of units is 10 and 20, and the number of interpolation points is 3, 6, and 10, satisfactory results have been obtained in other cases, but they are all time-consuming, the most are 49741.078125 s, which is 13.8170h. At the same time, the number of iterations is 500 when the number of units is 600 and the number of interpolation points is 10. Therefore, the key point method has advantages in terms of accuracy, but it takes a lot of time. If it is used to solve more complex systems, the time consumption will increase sharply, and it could even become intolerable.

3.5.2 Road condition two

The road surface profile of road condition two is shown in Figure 3-21. Vehicle speed $v = 24.384$ m/s, $t_\sigma = 0.125$ s, $\omega_i = 2\pi$ rad/s and 16π rad/s ($i = 1, 2, 3, 4$), suitable for two roads.

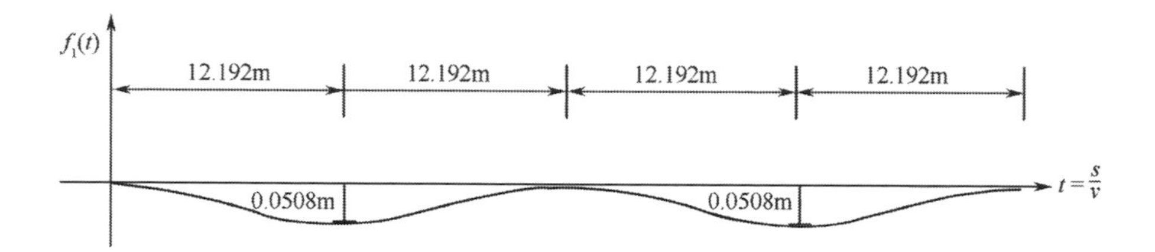

Figure 3-21 Road profile of road condition 2

Substitute the road condition functions $f_1(t)$ and $f_2(t)$ of road condition 2 into equation (3.3), and apply the spectral element method to solve them. The results are shown in Table 3.19–Table 3.23, and Figure 3-22–Figure 3-24.

Table 3.19 The best advantage of the GLL point is method numerical test (Road Condition 2, $\omega_l = 2\pi$ rad/s)

Number of interpolation points m		Number of units N_{el}								
		10	20	60	100	200	300	400	600	800
3	k_1	8756.3	8756.3	8756.3	8756.3	8756.3	8756.3	8756.300000	8756.300000	8756.300000
	k_2	35025.2	35025.2	35025.2	35025.2	35025.2	35025.2	35025.200000	35025.200000	35025.200000
	k_3	35025.2	35025.2	35025.2	35025.2	35025.2	35025.2	35025.200000	35025.200000	35025.200000
	c_1	8756.3	8756.3	8756.3	8756.3	8756.3	8756.3	8756.300000	8756.300000	8756.300000
	c_2	14010.08	14010.08	14010.08	14010.08	14010.08	14010.08	14010.080000	14010.080000	14010.080000
	c_3	12025.239	14010.08	12405.143	11886.817	12365.499	12006.002	12214.538067	12137.601981	12106.941020
	d	2.415443	2.389652	2.404458	2.404392	2.404489	2.404502	2.404555	2.404549	2.404546
6	k_1	8756.3	8756.3	8756.3	8756.3	8756.3	8756.3	8756.300000	8756.300000	8756.300000
	k_2	35025.2	35025.2	35025.2	35025.2	35025.2	35025.2	35025.200000	35025.200000	35025.200000
	k_3	35025.2	35025.2	35025.2	35025.2	35025.2	35025.2	35025.200000	35025.200000	35025.200000
	c_1	8756.3	8756.3	8756.3	8756.3	8756.3	8756.3	8756.300000	8756.300000	8756.300000
	c_2	14010.08	14010.08	14010.08	14010.08	14010.08	14010.08	14010.080000	14010.080000	14010.08000
	c_3	10706.198	11168.851	12469.341	11938.339	12161.315	12264.579	12205.037649	12214.828887	12191.425283
	d	2.399473	2.402319	2.404396	2.404432	2.404494	2.404543	2.404557	2.404555	2.404558

(Continued)

Number of interpolation points m		Number of units N_{el}								
		10	20	60	100	200	300	400	600	800
10	k_1	8756.3	8756.3	8756.3	8756.3	8756.3	8756.3	8756.300000	8756.300000	8756.300000
	k_2	35025.2	35025.2	35025.2	35025.2	35025.2	35025.2	35025.200000	35025.200000	35025.200000
	k_3	35025.2	35025.2	35025.2	35025.2	35025.2	35025.2	35025.200000	35025.200000	35025.200000
	c_1	8756.3	8756.3	8756.3	8756.3	8756.3	8756.3	8756.300000	8756.300000	8756.300000
	c_2	14010.08	14010.08	14010.08	14010.08	14010.08	14010.08	14010.080000	14010.080000	14010.080000
	c_3	12239.535	11353.901	11964.014	12087.739	12231.532	12160.188	12162.990727	12194.592986	12191.426661
	d	2.40455	2.402336	2.40447	2.404544	2.404552	2.404547	2.404558	2.404557	2.404558

Table 3.20 GLL point method numerical test CPU time consumption and number of iterations (Road Condition 2, ω_l = 2π rad/s)

Number of interpolation points m		Number of units N_{el}								
		10	20	60	100	200	300	400	600	800
3	t	19.921875	14.14062	54.265625	58.046875	193.60938	286.60938	214.687500	509.296875	415.531250
	n	95	60	95	67	115	125	75	130	82
6	t	25.625	44.53125	108.70313	153.71875	268.53125	631.10938	634.421875	880.828125	1571.234375
	n	97	102	95	85	77	121	84	85	90
10	t	31.046875	66.25	204.0625	342.73438	691.42188	1059.5781	2263.687500	2509.687500	4124.843750
	n	70	81	87	84	91	92	142	101	119

Table 3.21 The optimal value of the key point method numerical test (Road Condition 2, $\omega_l = 2\pi$ rad/s)

Number of interpolation points m		Number of units N_{el}								
		10	20	60	100	200	300	400	600	800
3	k_1	8756.300	8756.300	8756.300	8756.300	8756.300	8756.300	8756.300000	8756.300000	8756.300000
	k_2	35025.20	35025.20	35025.20	35025.20	35025.20	35025.20	35025.200000	35025.200000	35025.200000
	k_3	35025.20	35025.20	35025.20	35025.20	35025.20	35025.20	35025.200000	35025.200000	35025.200000
	c_1	6433.267	8464.265	8756.300	8756.300	8756.300	8756.300	8756.300000	8756.300000	8756.300000
	c_2	6968.144	9774.998	12976.43	13828.46	14010.08	14010.08	14010.080000	14010.080000	14010.080000
	c_3	5638.458	7520.940	9844.320	10472.06	11176.12	11706.51	12861.782791	11823.005867	12403.231285
	d	1.513167	2.012354	2.363677	2.388839	2.402354	2.404106	2.403714	2.404107	2.404460
6	k_1	8756.300	8756.300	8756.300	8756.300	8756.300	8756.300	8756.300000	8756.300000	8756.300000
	k_2	35025.20	35025.20	35025.20	35025.20	35025.20	35025.20	35025.200000	35025.200000	35025.200000
	k_3	35025.20	35025.20	35025.20	35025.20	35025.20	35025.20	35025.200000	35025.200000	35025.200000
	c_1	6721.638	8494.505	8756.300	8756.300	8756.300	8756.300	8756.300000	8756.300000	8756.300000
	c_2	6870.165	9765.838	12976.41	13828.46	14010.08	14010.08	14010.080000	14010.080000	14010.080000
	c_3	5483.822	7532.136	9844.383	10472.08	11176.14	11706.48	12861.769793	11823.005350	11985.009577
	d	1.530744	2.012644	2.363686	2.388839	2.402354	2.404106	2.403714	2.404107	2.404487

(Continued)

Number of interpolation points m		Number of units N_{el}								
		10	20	60	100	200	300	400	600	800
10	k_1	8756.300	8756.300	8756.300	8756.300	8756.300	8756.300	8756.300000	8756.300000	8756.300000
	k_2	35025.20	35025.20	35025.20	35025.20	35025.20	35025.20	35025.200000	35025.200000	35025.200000
	k_3	35025.20	35025.20	35025.20	35025.20	35025.20	35025.20	35025.200000	35025.200000	35025.200000
	c_1	6721.393	8494.492	8756.300	8756.300	8756.300	8756.300	8756.300000	8756.300000	8756.300000
	c_2	6870.271	9765.837	12976.41	13828.46	14010.08	14010.08	14010.080000	14010.080000	14010.080000
	c_3	5483.937	7532.144	9844.395	10472.07	11176.09	12864.02	12861.847829	11823.005543	11984.975934
	d	1.530754	2.012644	2.363686	2.388839	2.402354	2.403709	2.403714	2.404107	2.404487

Table 3.22 The CPU time and number of iterations of the key point method numerical test (Road Condition 2, $\omega_l = 2\pi$ rad/s)

Number of interpolation points m		Number of units N_{el}								
		10	20	60	100	200	300	400	600	800
3	t	356.656	563.468	1452.187	1219.937	3391.265	3411.593	6874.562500	8408.000000	7289.296875
	n	150	145	157	85	126	89	141	116	82
6	t	410.531	462.093	1286.687	1824.187	2954.875	2457.953	4558.109375	5869.468750	8295.234375
	n	171	112	130	117	100	58	86	72	79
10	t	455.328	602.281	1868.968	1626.234	2922.296	2696.046	5946.828125	7520.703125	14872.359375
	n	164	125	157	87	83	53	92	75	117

Table 3.23 Comparison of book results and literature results

Item	Results of this book			Literature [65] Results	
	Initial value	Optimal value		Initial value	Optimal value
		GLL point method	Key point method		
k_1	17512.6	8756.30	8756.30	17512.6	8756.30
k_2	52537.8	35025.20	35025.20	52537.8	35025.25
k_3	52537.8	35025.20	35025.20	52537.8	35025.25
c_1	1751.26	8756.30	8756.30	1751.26	1563.88
c_2	4378.15	14010.080	14010.08	4378.15	8041.79
c_3	4378.15	12214.82	11823.00	4378.15	6621.51
d (Objective function)	5.044	2.405	2.404	5.044	3.188

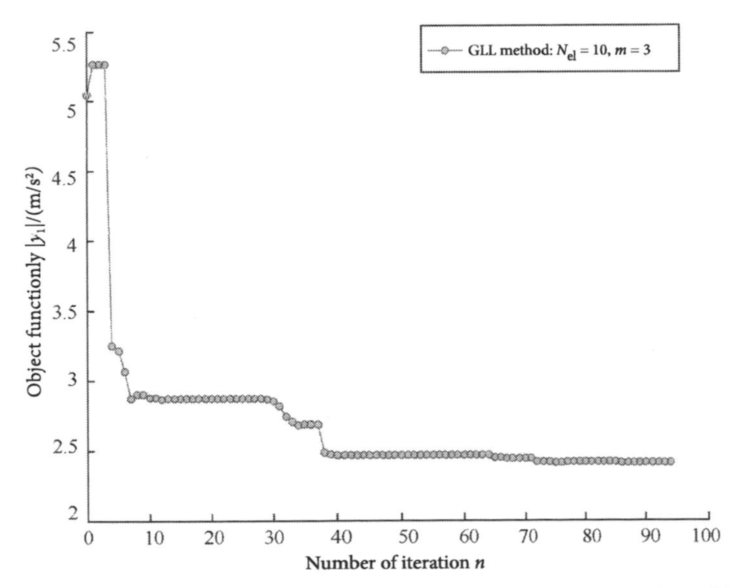

Figure 3-22 Using GLL point method for objective function iteration (road condition 2)

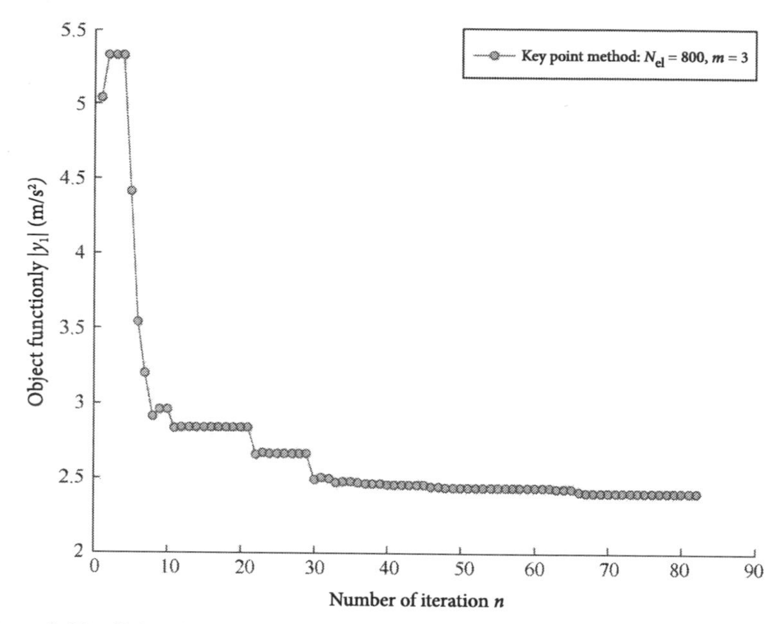

Figure 3-23 Using the key point method to iterate the objective function (road condition 2)

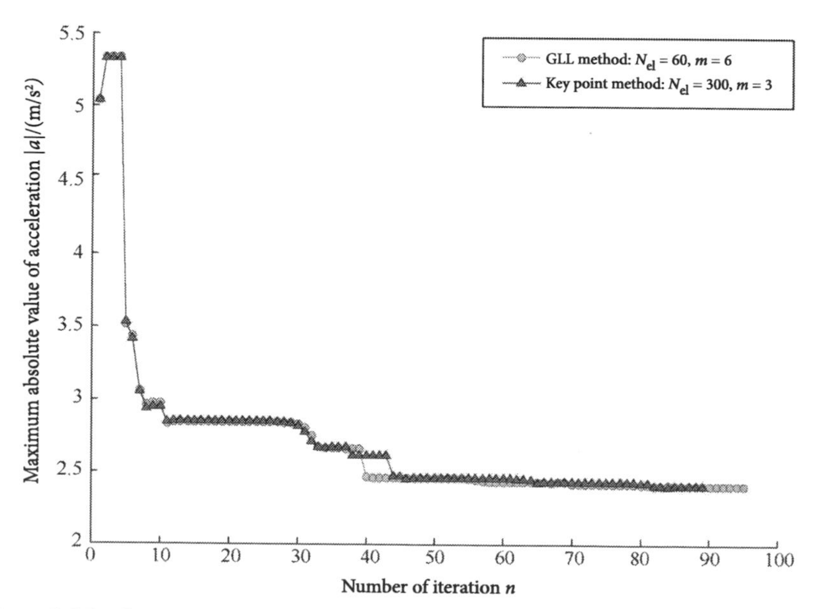

Figure 3-24 Comparison of the objective function iteration process of the two methods (road condition 2)

The dynamic response optimization design of the mechanical system has good application prospects. The dynamic response must satisfy a time-dependent differential equation. To obtain the optimal solution and satisfy the time-related constraints, it is required to obtain the response of the system over the entire time.

In this chapter, the spectral element method is used to calculate the response of the system over the entire time period, and the differential equations or equations are converted into algebraic equations, and then the sequential quadratic programming (SQP) optimization algorithm is used for optimization calculations. For time-related constraint processing, two methods are used, namely the key point method and the GLL point method. Through the analysis of linear single-degree-of-freedom system design problems, linear two-degree-of-freedom shock absorber design problems, and five-degree-of-freedom automotive suspension system design problems, the method verifies the feasibility of dynamic response optimization based on spectral element method. When dealing with time-related constraints and multi-degree-of-freedom dynamic problems, the GLL point method has superiority in time, although it requires more units. Although the key point method uses a small number of units, it is necessary to find the time-based optimal value for the high-order Lagrange function in each unit. The number of optimization searches is 2 × (the number of state variables + the number of combinations of state variables), then in the entire optimization process, at each iteration step, the number of times to find the time-based optimal value of the high-order Lagrange function is equal to the number of units × 2 (the number of state variables + the number of combinations of state variables). For example, if the number of units is 6, the number of state variables is 2, and the number of state variable combinations is 1, then the optimization is 36 times in 1 iteration. If there are 10 iterations in total, 360 times are required to optimize them, which is time-consuming. Therefore, the key point method has no advantage in time, but the key point method has a greater probability of finding the optimal value. Section 3.5 uses the GLL point method and the key point method to analyze and calculate the dynamic response optimization design problem of the five-degree-of-freedom automobile suspension system, which shows that in the multi-degree-of-freedom dynamic response optimization problem, increasing the number of elements and interpolation points can achieve better results. However, when the number of units and interpolation points increase to a certain number, a further increase will not affect the result, indicating that it has reached sufficient accuracy. The linear two-degree-of-freedom shock absorber design problem represents the simplest multi-degree-of-freedom dynamic design problem, the five-degree-of-freedom automotive suspension system design problem represents a more complex multi-degree-of-freedom dynamic design problem, other dynamic response optimization designs for mechanical systems problems, and so on.

For example, optimal design of dynamic response of a rectangular beam with variable cross-section subjected to vibration input at fixed end, means that the dynamic response optimization design of elastic beams are subjected to uniformly distributed transient dynamic loads on a vertical plane under different boundary conditions, etc. You can refer to the method in this chapter for analysis.

3.6 Summary of this chapter

This chapter studies the system dynamic response optimization algorithm based on the temporal spectral element method. It also deeply analyzes the discrete dynamic response of the system in the time domain, transforms the motion differential equations into algebraic equations, accurately solves their transient response, and uses the GLL point method and the key point method to deal with time constraints. Taking linear single-degree-of-freedom system design, linear two-degree-of-freedom shock absorber design, and five-degree-of-freedom automobile suspension system design as examples, artificial design variables are introduced, the advantages and disadvantages of the two methods for handling constraints are studied in detail, and the correctness of the system dynamic response optimization based on the temporal spectral element method is explained. These points can provide reference for further research on dynamic response optimization, such as sensitivity analysis of complex systems to improve the practicability of this method.

CHAPTER 4

Structural Dynamic Response Optimization Method Based on the Modal Superposition Method for Equivalent Static Loads of All Nodes

Almost all structures work under dynamic loads, so the various properties of the structure are functions of time. To improve the performance of the structure, dynamic response optimization is very necessary. The dynamic response optimization first requires the solution of structural transient dynamics. The problem of transient dynamics is that it is very time-consuming, and its optimization needs to repeatedly solve the objective function and constraint function. Besides, the direction of the structure under dynamic response optimization is often not convergent. Therefore, we should not only solve the transient dynamic problems quickly but also with other methods for structure dynamic response optimization.

Since the objective function and constraint function of dynamic response optimization are both functions of time and state variables, it is extremely expensive to conduct sensitivity analysis of design variables in dynamic response optimization. [118–120] In dynamic response optimization, it is almost impossible to deal directly with time-dependent functions, especially for large dynamic response optimization problems. [121–123] However, the development of structural static response optimization is relatively mature, and there has been a lot of research on replacing dynamic response optimization with structural static response optimization at home and abroad, among which the most widely adopted method is dynamic factor method. [124] Since dynamic factors are determined by a particular program or by the experience of the designer, it is difficult to find appropriate dynamic factors in this approach. The research team of G. J. Park proposed the equivalent static load method, [35, 115, 125–128] the main idea of which is to convert the dynamic load equivalent to a continuous static load and for it to act on the structure as a static multi-working condition for optimization. Domestic scholars have done a lot of research on equivalent static load method. The equivalent static load method based on gradient was proposed. [129]

It combines the structure's static linear optimization method and the maximum velocity descent method, based on the node displacement equivalence, and improves the convergence speed on the premise of ensuring the convergence of the algorithm. It has a great advantage in solving the optimization problem of large deformation structure with more design variables and more non-linear structures. The equivalent static load method and other methods were adopted to establish a simplified simulation model of the agitator deputy frame, and the performance analysis and optimization design of the agitator deputy frame were carried out, which not only ensured the calculation accuracy, but also improved the calculation efficiency and shortened the design cycle. [130] The equivalent static load method is adopted to optimize the crashworthiness design of the size and shape of the automobile's front-end structure. [131] It is applied to the whole structure, to optimize the objective function within the minimum quality department before the thickness of the main components of size and shape at the node coordinates design variables can invade the quantity and the reinforcement of the stamping process requirements. This is also done as constraint collision optimization design is carried out, and the equivalent static load method is adopted to nonlinear optimization problems for the transient collision condition linear static optimization problems. The constraint release was introduced as a boundary condition into topology optimization based on the equivalent static load method, using strain energy as an evaluation index for vehicle stiffness, and by introducing relative displacement as an evaluation index for component flexibility. [132] The similarities and differences of topology optimization by using release and single-point constraints as collision analysis models under vehicle head-on collision conditions are compared and studied, as well as the impact of different optimization objectives on the optimization results. The dynamic analysis of nonlinear flexible multi-body system is combined with the static optimization of linear structure based on the equivalent static load method, and it deals with the interaction of inertial forces of each component through the optimization of structural components under dynamic load. [133] The equivalent static load method was applied to optimize the connecting rod of vibrating screen and the auxiliary frame structure of the machine tool, and satisfactory results were obtained. [134–136] The structure dynamic optimization method under dynamic loading is proposed for using the traditional static optimization and dynamic optimization method of crankshaft connecting rod mechanism optimization design, which is based on the equivalent static load method and the nonlinear dynamics analysis of flexible multi-body system combined with linear static structure optimization. [137]

As such, dynamic optimization method is better than the static optimization method of conclusion. Li Ming et al. solved the problem of reliability topology optimization of interval parameter structures under dynamic response constraints with equivalent static load method, and gave new meaning to the equivalent static load. Mao Huping's research team [45, 64, 65, 139, 140] introduced the temporal spectral element method into dynamic response optimization, and identified key time points by virtue of the high accuracy of discrete interpolation by spectral element, thus, an equivalent static transformation method of volumetric strain energy was proposed and it achieved certain results.

In summary, the current equivalent static conversion in the equivalent static load method includes two types: a) equivalent static conversion at key time points; b) equivalent static conversion at all time points. Between them, the load includes the equivalent static load of some nodes and the equivalent static load of all nodes. The equivalent static conversion at all time points is very time-consuming, and the equivalent static load solution model of some nodes is uncertain. The identification of key time points will lead to the uncertainty of the upper and lower boundary values of the equivalent static load, and the arbitrariness of the initial values. It is therefore necessary to solve the structural static analysis problems multiple times during the iteration process, resulting in nested optimization. The reliability and efficiency of static conversion can become worse. Therefore, this chapter adopts new ideas to study related issues from the perspective of getting rid of nesting optimization and improving reliability and efficiency.

4.1 Modal superposition method

For the undamped vibration structure, the motion differential equation of the finite element method is:

$$\boldsymbol{M}\ddot{\boldsymbol{d}} + \boldsymbol{K}\boldsymbol{d} = \boldsymbol{F} \tag{4.1}$$

Where $\boldsymbol{M}$ is the mass matrix; $\boldsymbol{K}$ is the stiffness matrix; $\boldsymbol{F}$ is the dynamic load vector; d is a coordinate change. All of them are defined by using eigenvectors for dynamic displacement vector:

$$\boldsymbol{d} = \boldsymbol{\Phi}\boldsymbol{z} \tag{4.2}$$

Where, $\boldsymbol{\Phi} = \{\boldsymbol{\phi}_1, \boldsymbol{\phi}_2, \cdots, \boldsymbol{\phi}_n\}$ is the modal matrix, where, $\boldsymbol{\phi}_i = \{u_{1i}, u_{2i}, \cdots, u_{ni}\}^{\mathrm{T}}$ is the eigenvector; z is the modal coordinate vector. Substitute Equation (4.2) into Equation (4.1) and get:

$$\ddot{\boldsymbol{z}} = \boldsymbol{\Omega}^2\boldsymbol{z} = \boldsymbol{Q} \tag{4.3}$$

Where, $\boldsymbol{\Omega}^2=\mathrm{diag}\left[\omega_1^2,\omega_2^2,\cdots,\omega_n^2\right]$ is the natural frequency matrix; $\boldsymbol{Q}=\boldsymbol{\Phi}^{\mathrm{T}}\boldsymbol{F}$ is the modal force matrix.

Equation (4.3) is a set of decoupled equations, which can be written as:

$$\ddot{z}_i+\omega_i^2 z_i=f_i \quad (i=1,2,\cdots,n) \tag{4.4}$$

Where, f_i is the ith row of the modal force matrix.

The solution of Equation (4.4) can be obtained by Duhamel integral:

$$\begin{aligned} z_i(t)=z_i(t_0)\cos(\omega_i t)+\frac{\dot{z}_i(t_0)}{\omega_i}\sin(\omega_i t)+ \\ \frac{1}{\omega_i}\int_0^t \sin(\omega_i (t-\tau)f_i(\tau))\mathrm{d}\tau \end{aligned} \tag{4.5}$$

When the response in modal coordinates is obtained, the response in the actual coordinate system can be superimposed:

$$d_i(t)=\sum_{j=1}^{n}\phi_{ij}z_j(t) \quad (i=1,2,\cdots,n) \tag{4.6}$$

4.2 Equivalent static load method

In the linear static analysis, the static load analysis of the structure produces the same response field as the nonlinear dynamic analysis of the structure at that corresponding moment. There are two methods for equivalent static conversion of dynamic load from the point of static load. The first is to apply equivalent static load on some nodes, and the second is to apply equivalent static loads at all nodes. The former load size load point, such as randomness, usually through trial calculation can be determined. Although the latter can accurately calculate the equivalent static load at each moment, multi-condition handling is a tricky problem for further optimization. The principle of the equivalent static load method is described as follows.

As shown in Figure 4-1, the calculation step in the transient dynamic analysis is $N+1$, and each step is equivalent to a static working condition, and the displacement field under the equivalent static load is equal to the displacement field under the dynamic load at the corresponding moment.

Based on this principle and combined with the modal superposition method, the equivalent static load $f_{eo}, f_{e1}, f_{e2}, \cdots, f_{en}$ corresponding to each time point can be obtained.

However, when the calculation step is relatively large, the equivalent static load is also relatively large, increasing optimization calculation.

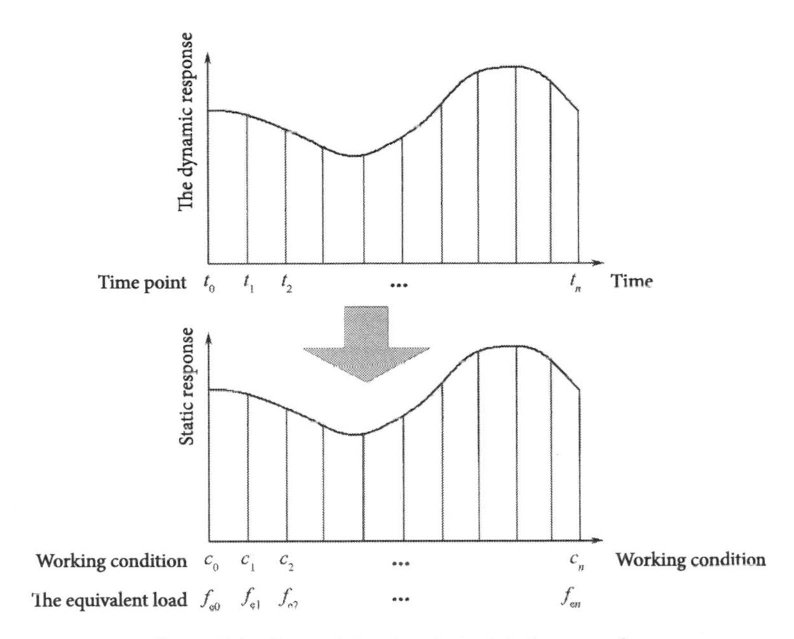

Figure 4-1 Dynamic load equivalent static conversion

To cover all possibilities, you can increase the time point, but the increase in time point incurs a large computational cost. In the selected time point, when the design parameter changes, there will be no extreme response at some time points, so this part of the time point can be eliminated completely. If we optimize the equivalent static load at each time point as a single working condition, it becomes a great challenge to the optimization algorithm, and sometimes the results may diverge. For linear systems, a large displacement indicates a large load. The displacements we care about are often the displacements of the system in one direction, so when the system acquires large displacements in one direction, its displacements in other directions may be small. In the optimization process, the maximum displacement time point may change locally with the change of parameters. Therefore, we cannot only take the maximum displacement time point in a direction we care about as the key point. The key time points can be obtained by making the interpolation derivative of the spectral element's discrete Lagrange equal to zero, and then the set of key time points can be formed with its two adjacent GLL points, as shown in Figure 4-2. Where p_i, p_{j1}, p_{j2} is the solution and where the interpolation derivative of the spectral element's discrete Lagrange is equal to zero; p_{j1}, p_{j2} are the two closest GLL points to p_i, p_{j1}, p_{j2} are the two closest GLL points to p_{j1}.

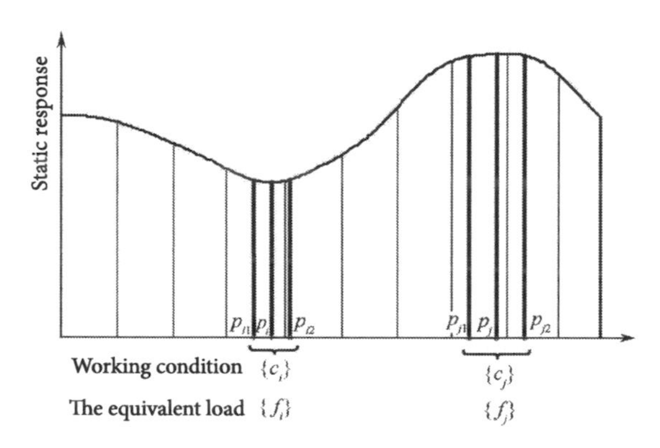

Figure 4-2 Key time point set diagram

The equivalent static load can be obtained by formula (4.7).

$$\boldsymbol{s} = \boldsymbol{Kd}(t_c) \tag{4.7}$$

Where, s represents the exact equivalent static load, and all degrees of freedom act on one load; can be obtained in two ways. One is the transient dynamic analysis and the other is the modal superposition method. Equation (4.7) not only shows that the equivalent static load does exist but also shows that it has an accurate value. To use the modal superposition method, the modal matrix and stiffness matrix of Equation (4.7) are extended. Substitute Equation (4.2) into Equation (4.7) and get:

$$\boldsymbol{d}(t_c) = \boldsymbol{\Phi}(\boldsymbol{\Phi}^{\mathrm{T}}\boldsymbol{K}\boldsymbol{\Phi})^{-1}\boldsymbol{\Phi}^{\mathrm{T}}\boldsymbol{s} \tag{4.8}$$

The load vectors in Equation (4.7) and Equation (4.8) are consistent. Equation (4.7) requires transient analysis, while Equation (4.8) only requires modal analysis and other calculations.

4.3 Critical time point set

Lagrange interpolation is carried out on GLL points. After obtaining a high-precision interpolation function, the first derivative of this interpolation function related to time is calculated and made equal to zero, then the key time point can be obtained.

The interpolation function is:

$$L(t)=\sum_{e=1}^{q}\sum_{j=1}^{k}x_j^e p_j^e(t) \tag{4.9}$$

Where, $p_j^e(t)=\prod_{i\in I_j}\frac{t-t_i}{t_j-t_i}$, where, $I_j=\{1,2,\cdots,\hat{j},\cdots,k\}$, $t_0\leqslant t\leqslant t_1$, (is the start time of simulation, usually 0; t_1 is the end time of simulation); q is the number of discrete elements in the solution space; k is the discrete GLL points of each cell; x_j^e is the dynamic response corresponding to the jth point GLL of element e (such as displacement stress).

By the differential theorem it will be:

$$\frac{\mathrm{d}L(t)}{\mathrm{d}t}=\sum_{e=1}^{q}\sum_{j=1}^{k}x_j^e\frac{\mathrm{d}p_j^e(t)}{\mathrm{d}t} \tag{4.10}$$

Where, $\frac{\mathrm{d}p_j^e(t)}{\mathrm{d}t}=\sum_{r=1}^{k}\prod_{i\in I_j}\frac{t-t_i}{(t_j-t_i)(t-t_r)}$. The critical time point can be obtained by solving $\frac{\mathrm{d}L(t)}{\mathrm{d}t}=0$, and then the left and right GLL points closest to the critical time point can be obtained by comparison. The key time point and these two GLL points can constitute the key time point set.

4.4 Method flow

The specific implementation process of this chapter is as follows:

Step 1: Modal superposition transient dynamics analysis.

Step 2: Determining the maximum stress element.

Step 3: Calculating the stress corresponding to the GLL point of the maximum stress element in discrete time with the spectral element.

Step 4: Adopting Equation (4.9) for Lagrange interpolation and solving Equation (4.10) to obtain the key time points, and then finding the adjacent left and right GLL points to form the set of key time points.

Step 5: Calculating the equivalent static load of all nodes in the critical time point set at each time point through Equation (4.8), and form the equivalent static load vector set.

Step 6: Applying the set of equivalent static load vectors solved in Step 5 to the corresponding node, and then conducting static optimization.

The set of equivalent static load vectors corresponding to the set of critical time points can be analyzed as multiple loading conditions. The static analysis is carried out for each working condition respectively, and then the unit stress values under each working condition are saved. After all working conditions are calculated, the number of elements and the unit stress values form a two-dimensional array, among which the unit stress values will form the maximum envelope values.

Step 7: Checking whether the static optimization converges. If not, updating the design variable and returning step 1; otherwise, ending the optimization.

4.5 Example analysis

4.5.1 The 124-bar truss structure size optimization

The 124-bar truss in this example has 49 hinges and 94 degrees of freedom [as shown in Figure 4-3 (a)]. The elastic modulus is $E = 207$ Gpa, Poisson's ratio is $v = 0.3$, density $\rho = 7850$ kg/m^3, the cross area of the bar is 0.645×10^{-4} m^2. The dynamic load is half-sine function [as shown in Figure 4-3 (b)]. The same dynamic load is applied upward on the positive X-axis of nodes 1, 20, 19, 18, 17, 16, 15, and upward on the negative Y-axis of nodes 1, 2, 3, 4, and 5.

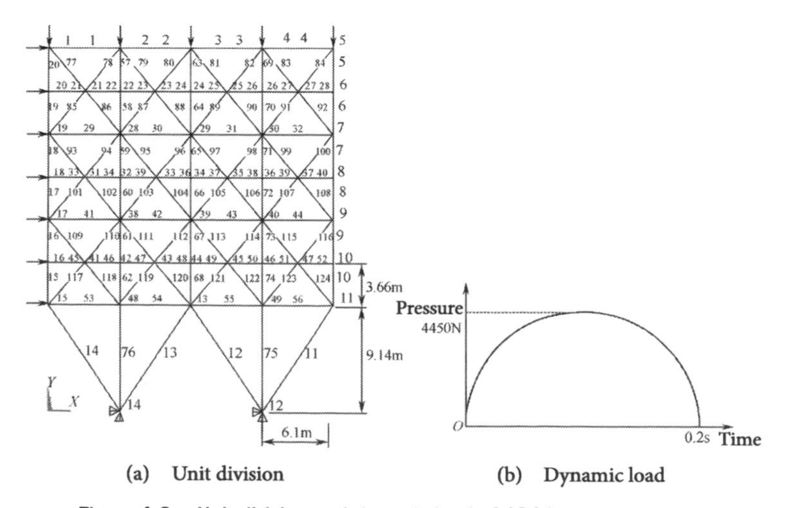

(a) Unit division (b) Dynamic load

Figure 4-3 Unit division and dynamic load of 124 bar truss structure

To reduce the challenge of optimization, 124 rods were divided into 6 categories (see Table 4.1): left-leaning to right-leaning horizontal long rods, horizontal short rods, vertical short rods bracket rods, including 11 design variables. The objective of the optimization is the lightest mass and the constraint is stress constraint, that is, the maximum stress is less than 147.7 mpa.

The initial value of each design variable is 645 cm^2.

Table 4.1 Variable grouping of 124-bar truss

Design variable	Corresponding unit
X_{DV1} (Tilt to the left, 24 units)	77, 79, 81, 83, 86, 88, 90, 92, 93, 95, 97, 99, 102, 104, 106, 108, 109, 111, 113, 115, 118, 120, 122, 124
X_{DV2} (Tilt to the right, 24 units)	78, 80, 82, 84, 85, 87, 89, 91, 94, 96, 98, 100, 101, 103, 105, 107, 110, 112, 114, 116, 117, 119, 121, 123
X_{DV3} (Horizontal long pole, 16 units)	1–4, 29–32, 41–44, 53–56
X_{DV4} (Horizontal short pole, 24 units)	21–28, 33–40, 45–52
X_{DV5} (Vertical short pole, 30 units)	5–10, 15–20, 57–74
X_{DV6}–X_{DV11} (Bracket rod, 6 units)	X_{DV6}: 11 X_{DV7}: 12 X_{DV8}: 13 X_{DV9}: 14 X_{DV10}: 75 X_{DV11}: 76

Figure 4-4 shows the comparison of all node displacements under equivalent static load (ESL) at key time point 0.101726470521158 s and all node displacements under dynamic load (DYAN) at that time. As can be seen from the figure, the two displacements coincide well, and UX displacement is significantly greater than UY displacement. Figure 4-5 is the relative error of UX displacement of nodes under equivalent static load at this critical time point. As can be seen from the figure, the maximum relative error of UX displacement is 2.953%. In Figure 4-5, node 12 and 14 are displacement constraint points, so the results are discontinuous. Because space is limited, the relative displacement error of other moments will not be described here.

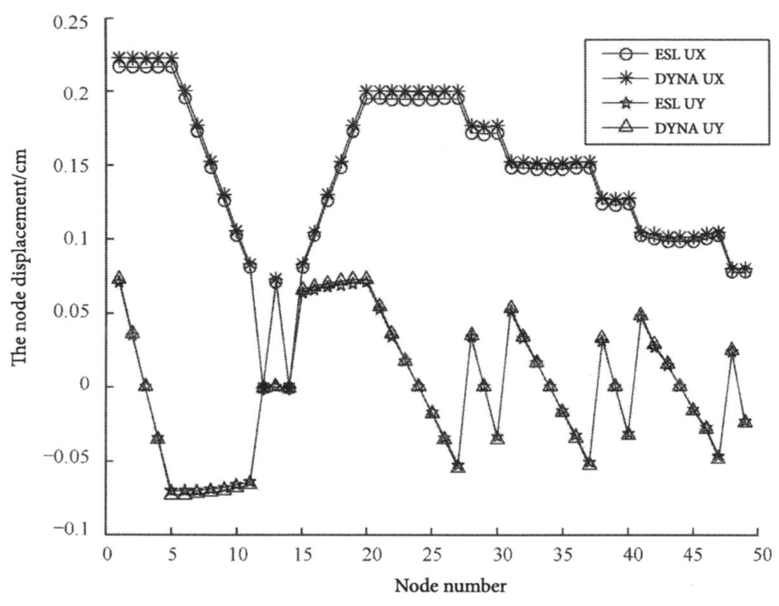

Figure 4-4 Comparison of the displacement of all nodes under the equivalent static load (ESL) corresponding to 0.101726470521158 s at the critical time point and the displacement of all nodes under the action of the dynamic load (DYAN) at that moment

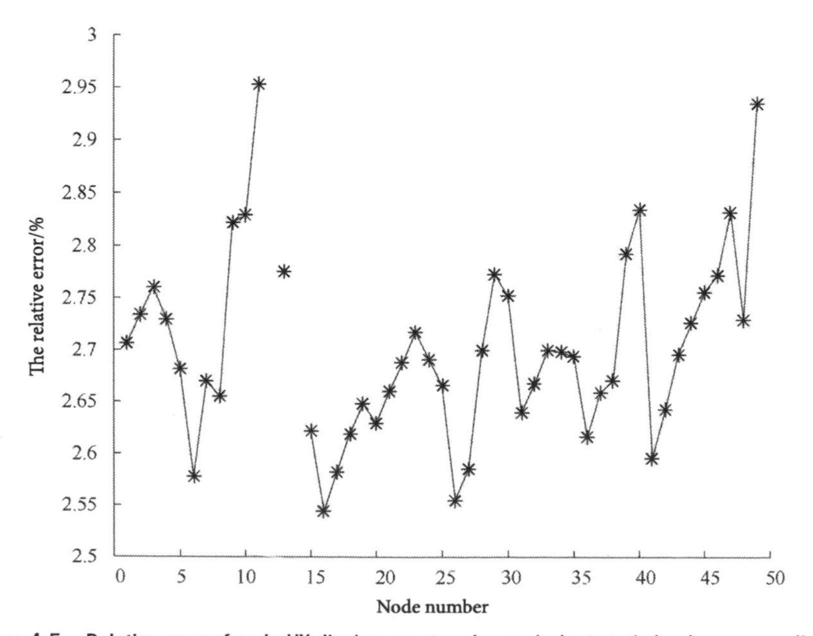

Figure 4-5 Relative error of node UX displacement under equivalent static load corresponding to 0.101726470521158 s at the critical time point

Under the action of dynamic load, there is an equivalent static load at all time points and degrees of freedom at all nodes. Figure 4-6 shows the equivalent static load at all time points corresponding to the *X* and *Y* degrees of freedom at nodes 1 and 5. It can be seen from the figure that the trend of the equivalent static load is consistent with the dynamic load. Even if the dynamic load is removed, the equivalent static load still exists. It can be seen from Figure 4-7 that when the dynamic load reaches the maximum in 0.1 s, the stress of the dangerous unit 13 does not reach the maximum. The stress reached the maximum at 0.101726470521158 s, indicating that the dynamic effect produced stress hysteresis. When comparing Figure 4-6 with Figure 4-7, it is not difficult to see that Figure 4-7 is obtained by making equation (4.10) equal to zero.

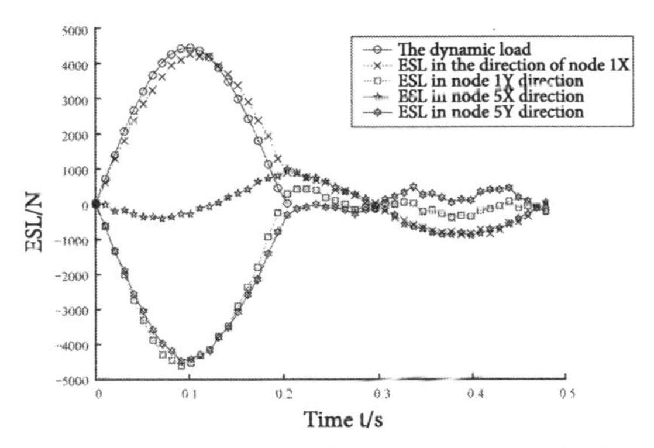

Figure 4-6 Equivalent static loads at all time points corresponding to the X and Y degrees of freedom of nodes 1 and 5

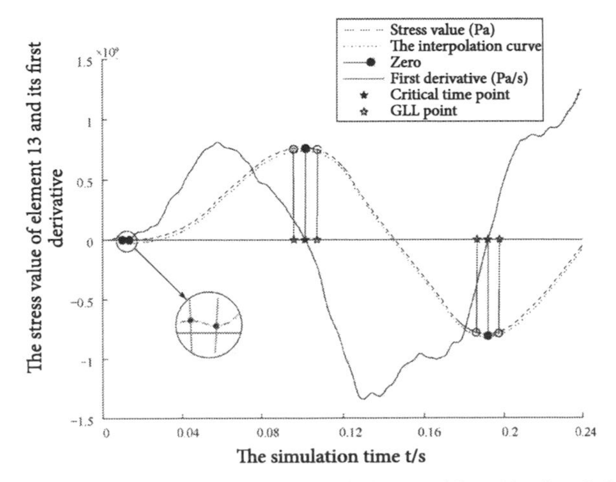

Figure 4-7 The stress value of hazard element 13 and its first derivative

It can be seen from Table 4.2 and Figure 4-8 that the critical time point set method and all time point method can achieve good convergence, but the former reaches the optimal target value of 491.82991700 kg after 210 iterations and takes 120.9652 min, while the latter reaches the optimal target value of 524.39984650 kg and takes 1972.6639 min after 217 iterations. The latter took 16.3077 times longer than the former, resulting in a reduction of 78.0747% and a reduction of 76.6228%. Table 4.3 shows the comparison of the 124 truss optimal design variables obtained by the two methods.

Table 4-2 Comparison of time-consuming and optimal target value of 124-bar truss obtained by key time point set method and all time point method

Method	Time consuming	Optimal target value
The initial value		2243.2100 kg
Critical point set method	120.9652 min	491.82991700 kg
All time points method	1972.6639 min	524.39984650 kg

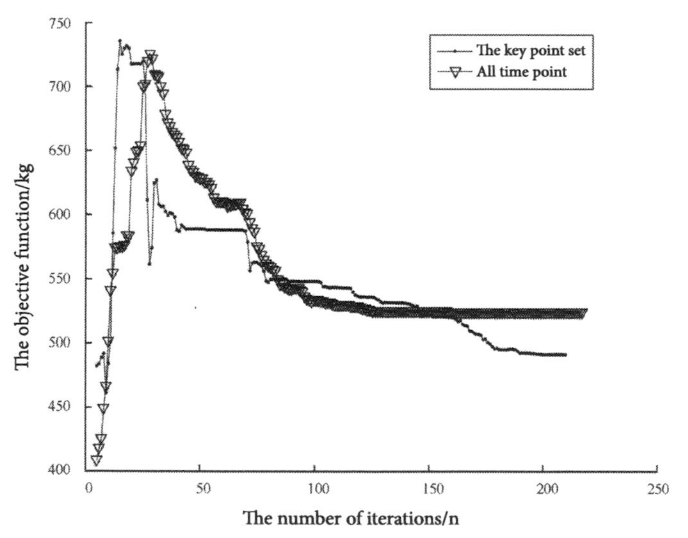

Figure 4-8 Optimization comparison between key time point set method and all time point method (124-bar truss)

Table 4.3 Comparison of the optimal design variables of the 124-bar truss obtained by the key time point set method and the all time point method (Unit: cm²)

The design variables	Critical point set method	All time points method
X_{DV1}	93.6142682200	104.563531176275
X_{DV2}	93.4741078900	106.100841084417
X_{DV3}	162.985520320	157.462708329796
X_{DV4}	10.6726181100	30.1563613576239
X_{DV5}	65.0455224200	71.5098397135004
X_{DV6}	228.745782860	242.272270401740
X_{DV7}	441.610451730	397.237977772873
X_{DV8}	380.350871260	429.337221064011
X_{DV9}	274.622532290	209.877297778638
X_{DV10}	152.999431070	176.382418625542
X_{DV11}	160.706131190	177.636602101239

4.5.2 Hybrid optimization of size and shape of 18-bar truss structure

For the 18-bar truss structure in Figure 4-9, the semi-sinusoidal dynamic ad is applied to its nodes 1, 2, 4, 6 and 8. Elastic modulus $E = 69$ Gpa, density $\rho = 2765$ kg/m³, Poisson's ratio $\mu = 0.3$.

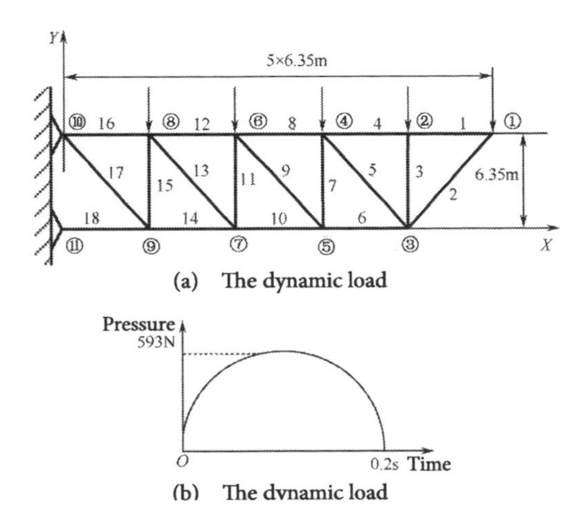

Figure 4-9 18 The structural element division and dynamic load of bar truss

We have to optimize both the size – the cross-sectional area of 18 rods – and the shape of the truss structure. If the cross-sectional area of each rod is taken as a design variable, the optimization will be greatly challenged. Moreover, it is necessary to consider the node coordinates to describe the shape of the truss structure. Therefore, the design variables are divided into size variables and shape variables, and then 18 rods are grouped, the cross-sectional area of each group is taken as a design variable, and the X and Y coordinates of node 3, 5, 7 and 9 are used to describe the shape of the truss structure (see Table 4.4).

Table 4.4 18 Grouping of Variables of Bar Truss

Design variable	X_{DV1} (Upper level)	X_{DV2} (Below level)	X_{DV3} (Upright)	X_{DV4} (Diagonal)
Corresponding unit	1, 4, 8, 12, 16	2, 6, 10, 14, 18	3, 7, 11, 15	5, 9, 13, 17
design variable	X_{DV5}–X_{DV6}	X_{DV7}–X_{DV8}	X_{DV9}–X_{DV10}	X_{DV11}–X_{DV12}
Corresponding node coordinates	(X_3, Y_3)	(X_5, Y_5)	(X_7, Y_7)	(X_9, Y_9)

The initial values of size variables are 8400 mm^2, and the initial values of shape variables are 0. The objective of optimization is to minimize mass, and two constraints are maximum stress constraint and maximum displacement constraint. The optimization model is:

$$\begin{cases} \min\limits_{X} \text{mass} \\ \text{s.t.} \left|\sigma\right|_{\max} \leqslant 177.9\,\text{MPa} \\ \text{s.t.} \left|\sigma\right|_{\max} \leqslant 28.4\,\text{cm} \end{cases}$$

As can be seen from Figure 4-10, under the action of the semi-sinusoidal dynamic load, the equivalent static load of each node is oscillating, and compared with the dynamic load, the moment when it reaches the extreme value is either comes earlier or later, indicating that the equivalent static load is real under the action of the dynamic load.

Figure 4-11 shows the comparison of all node displacements corresponding to the equivalent static load at the key time point 0.121677775449288 s with all node displacements under the dynamic load at that time. It can be seen from the figure that the two results are in good agreement. Figure 4-12 shows the relative error of UX, UY displacement of nodes under static load at this moment, it can be seen from the figure that the maximum relative error of UX displacement is 3.209%, and the maximum relative error of UY displacement is 0.893%.

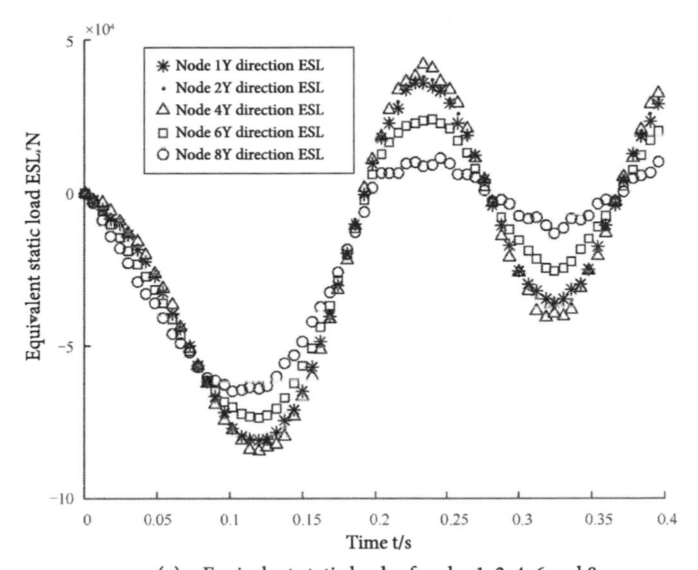

(a) Equivalent static loads of nodes 1, 2, 4, 6 and 8

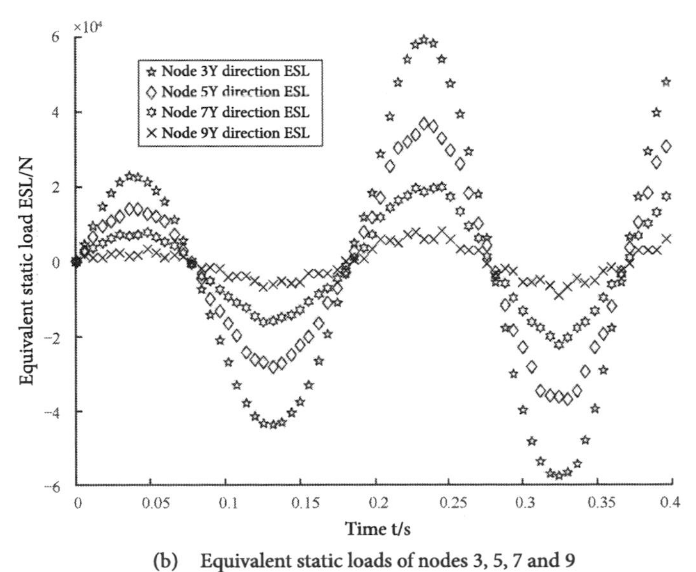

(b) Equivalent static loads of nodes 3, 5, 7 and 9

Figure 4-10 Equivalent static load of nodes in time history

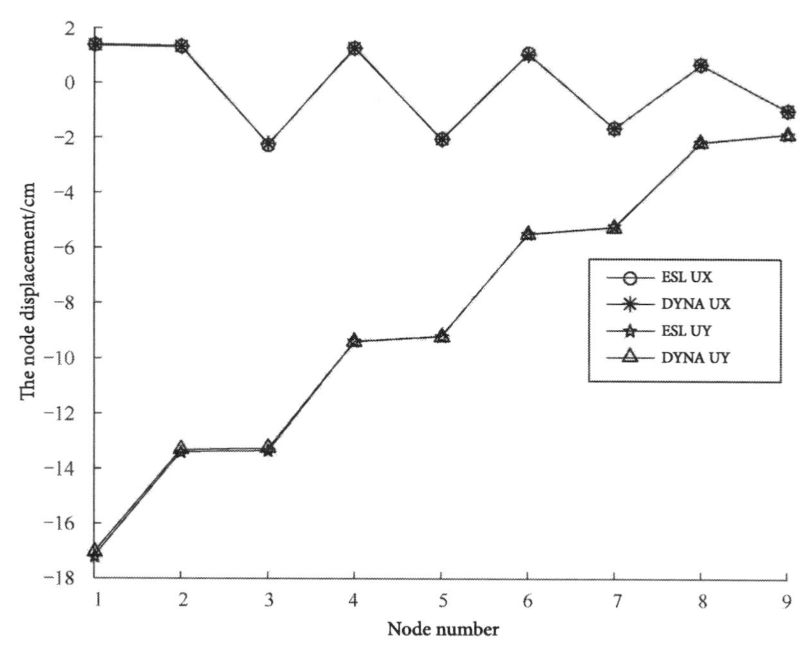

Figure 4-11 Comparison of all nodal displacements under the action of equivalent static load at the critical time point 0.121677775449288 s and all nodal displacements under the action of dynamic load at that moment

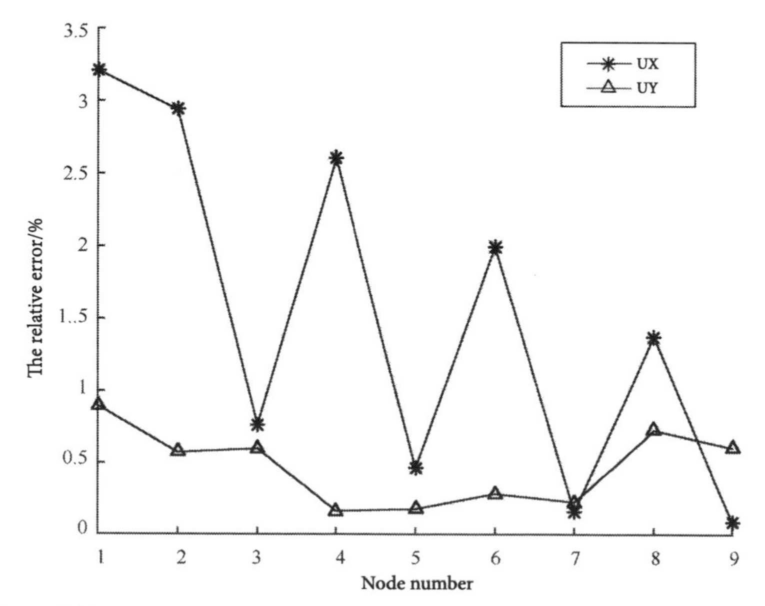

Figure 4-12 Relative error of node UX and UY displacement under equivalent static load corresponding to 0.121677775449288 s at critical time point

Figure 4-13 shows the stress value of hazard element two and its first derivative. Three extreme points can be obtained by making the first derivative equal to zero. The one with the largest absolute value is selected as the critical time point, and constitutes the critical time point set with the adjacent left and right GLL points.

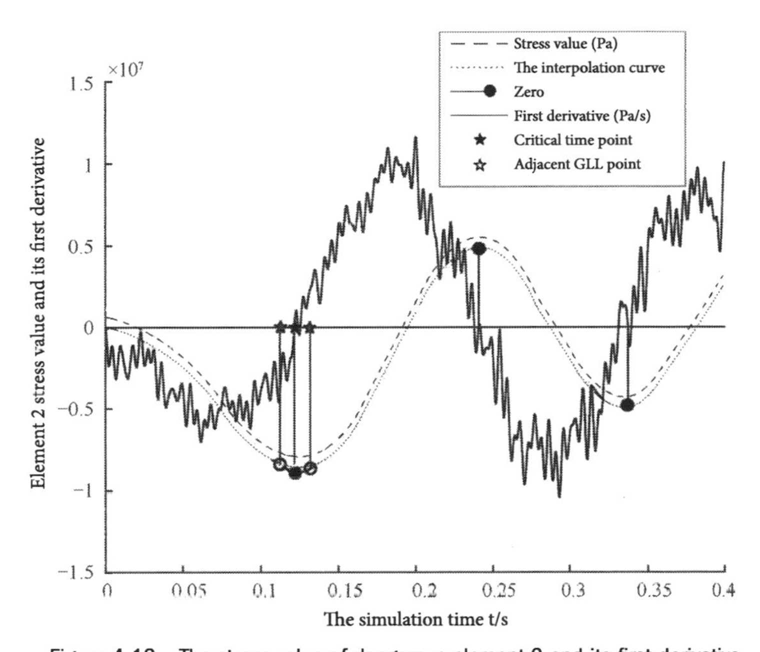

Figure 4-13 The stress value of dangerous element 2 and its first derivative

The key time point set method and all time point method were used for optimization, and the results shown in Table 4.5, Table 4.6 and Figure 4-14 and Figure 4-15 were finally obtained. As shown in Figure 4-15, for the two methods, the geometric positions of nodes 1, 2, 4, 6, 8, 10, 11 and cells 1, 4, 8, 12, 16 in the figure remain unchanged, while the positions of other nodes change. Among them, the shape control points obtained by all time point methods are nodes 3, 5, 7 and 9, while the shape control points obtained by the critical time point set method are nodes 12–15, and the attitude of their corresponding middle rod changes, therefore, the shapes obtained by the two methods are significantly different, see X_{DV5}–X_{DV12} in Table 4.5 for specific data (this is assuming that the original coordinates of shape design variable X_{DV5}–X_{DV12} are taken as the origin and the optimization result is the relative distance from the original coordinates). From the perspective of engineering experience, the results obtained by the critical time point set method are more reliable.

Table 4.5 Comparison of 18-bar truss optimization results obtained by the key time point set method and all time point methods (Unit: cm^2)

The design variables	Critical point set method	All time points method
X_{DV1}	7670.00941882803	9045.99833623767
X_{DV2}	10261.3757183491	9855.38007354326
X_{DV3}	2826.20567659979	2428.43674911552
X_{DV4}	4656.08514566531	3914.27947441947
X_{DV5}	−2.62821142871100	−4.89134104632723
X_{DV6}	2.12342718172536	2.32787313012177
X_{DV7}	−0.84426690690217	−3.0872497751829
X_{DV8}	1.26857341292592	1.46441658162896
X_{DV9}	−0.90146864304572	−1.59842320557510
X_{DV10}	0.112760509940237	0.632538793959340
X_{DV11}	−1.27256777545745	−2.20654084056575
X_{DV12}	−0.27638200133142	−0.125095356332809

Table 4.6 Comparison of 18-bar truss optimization time consumption and optimal target value obtained by using the key time point set method and all time point methods

Method	Optimize the time-consuming	Optimal target value
The initial value		2960.183443 kg
Critical point set method	70.339567 sec	2170.8819 kg
All time points method	878.076883 sec	2124.3312 kg

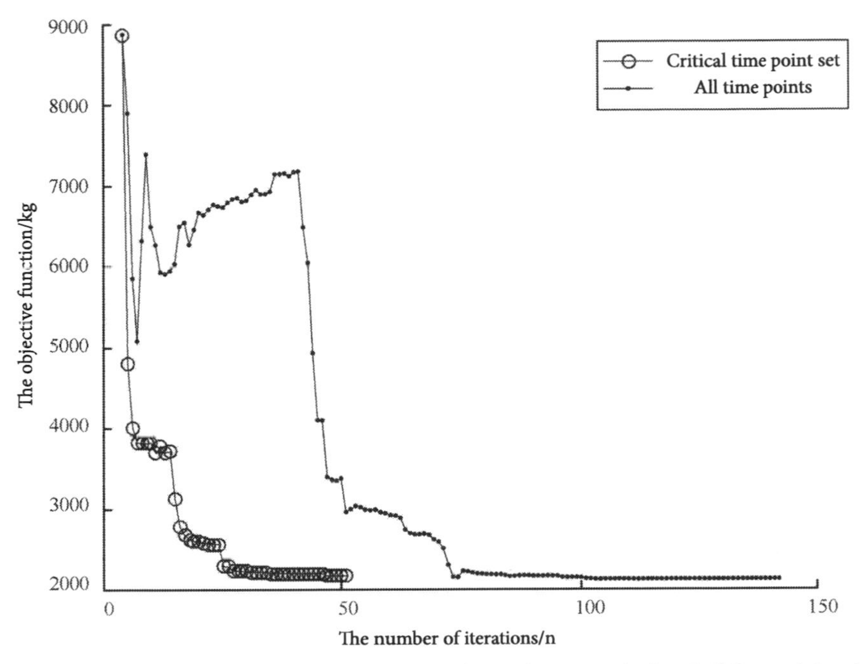

Figure 4-14 Optimization comparison between key time point set method and all time point method (18-bar truss)

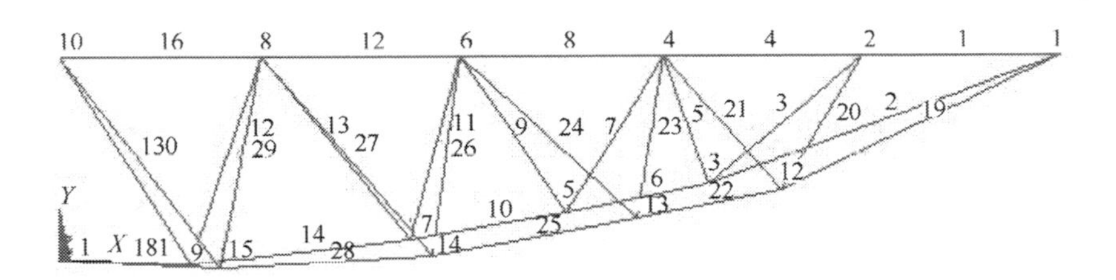

Figure 4-15 Optimization results of all nodes obtained by two methods under equivalent static load

It can be seen from Table 4.6 and Figure 4-14 that after 23 iterations of the critical time point set method, the objective function value has reached 2,234.9604 kg, which is very close to the optimal target value of 2,170.8819 kg. However, after 23 iterations of all-time point methods, the objective function value was 6772.0975 kg, which was far from the optimal target value, and after 75 iterations, the objective function value reached 2231.0834 kg. Comparatively, the critical time point set method could converge rapidly. From the perspective of optimization time, all-time point methods take 878.076883 min, while the critical time point set method only takes 70.339567 min, the former is 12.4834 times of the latter. From the average iteration step time, all-time point methods were 6.1836 min/step, and the key time point set method was 1.3792 min/step. The step time of the former was 4.4835 times that of the latter. In terms of mass reduction, the total time point method decreased by 28.2365% and the critical time point set method decreased by 26.6639%. For engineering, both of them are close to the optimal solution.

4.6 Summary of this chapter

This chapter discussed the application of the modal superposition method for all time points, the equivalent static load of all nodes, and it extracted the stress of the most dangerous unit by GLL interpolation to obtain high accuracy of the interpolation function to make its first derivative equal to zero, get the key point in time, and make sure the key point and adjacent around two GLL points constituted the key time point set. The equivalent static loads obtained by the critical time point set method and all the time point methods were applied to the structure as multi-working loads, and the structure was optimized to obtain the following conclusions:

1) The modal superposition method can obtain the equivalent static load at any time point only through modal analysis, avoiding the transient dynamic analysis.
2) By making the first derivative of the discrete Lagrange interpolation function time equal to zero, the extreme value point of the interpolation function can be obtained. By setting the threshold value, the critical time point can be obtained through simple filtering, which applies to all extreme value points.
3) The key time point set method eliminates the process of solving the equivalent static load by optimizing the nesting. It eliminates all the irrelevant time points in the time point method and greatly improves the computational efficiency. For example, in the optimization of the size of 124 truss structures, the key time point set method takes 120.9652 min, while all-time point methods take 1972.6639 min. In the hybrid optimization of the size and shape of 18-bar truss structure, the critical time point set method took 70.339567 min, while all-time point methods took 878.076883 min.

CHAPTER 5

Continuous Structural Optimization Method Based on Local Feature Sub-structure Method

It is very difficult to optimize large-scale continuous structures, which are mainly manifested in two aspects: a) The finite element analysis of large-scale continuous structures is very time-consuming, resulting in low efficiency in the optimization process; b) The parameterization of continuous structures is also more difficult. For simple elements such as rod elements, plane elements, beam elements, elements of the cross-sectional area, elements' in the thickness, and beam cross-sectional areas are often used as design variables, and it is easy to parameterize them, but continuous structures do not have these characteristics. According to the advantages of the sub-structure method and the characteristics of each sub-function in the optimization process, the optimization method of the continuous structure needs to be reconsidered.

When analyzing large continuous structures or complex structures, the general finite element method will encounter problems such as insufficient computer capacity or excessive machine time. To overcome such difficulties, the displacement substructure method can be used to study the plane problem. Under the condition of satisfying the equilibrium conditions and compatibility conditions at the nodes in the substructure, it integrates the forces and deformations of each substructure and divides the huge original structure into several substructures for calculation. [141] The main mode superposition method is a special Ritz vector, which only reflects the dynamic characteristics of the structure itself, and has nothing to do with the dynamic load on the structure. Therefore, when determining the structure's response to external dynamic loads, it is impossible to know in advance which Ritz vectors are the main contributions, nor can it be determined how many Ritz vectors should be used for superposition. For this reason, Wilson proposed the Ritz vector direct superposition method, in which the choice of the Ritz vector is not only related to the dynamic characteristics of the structure itself but also related to the spatial distribution of the load on the structure. On this basis, Lou Menglin [142] proposed a static substructure method suitable for structural dynamic analysis. In the finite element analysis of steam turbine components, we often encounter the problem of a high order of linear algebraic equations formed by finite elements. For example, when analyzing the strength of a group of blades, several thousand-order equations can be obtained. Therefore, first this book considers the computer's capacity and calculation speed. Zheng Xinyuan [143] studied the static substructure method and dynamic substructure method suitable for large-scale structural problems. Li Yuanke et al. [144] adopted the idea of sub-structure to simplify the high-order stiffness equation into a low-order aggregation equation and then introduced the contact boundary conditions from the relationship between the displacement and the force of the contact point, and finally performed a finite element analysis on the rolling bearing. In the decomposition method of the sensitivity solution, people are more concerned about the displacement coupling information between the sub-problems, because the stress constraint coupling between the sub-problems depends on the displacement coupling. Based on this point, Zhao Wenzhong et al. [145] proceeded from the concept of substructure in structural analysis, proposed a substructure method based on sensitivity analysis, and conducted in-depth research on substructure technology. Topological optimization takes the unit as the topological variable. For large continuous structures, the topological variable increases with the increase of the number of units, which makes its optimization difficult. With the help of the idea of sub-structures, a complex structure can be decomposed into multiple sub-structures, and then optimized separately, finally achieving the goal of optimizing the overall structure. [146] For example, to obtain the force transmission path of different internal force load requirements, first, use the sub-structure method to separate the structure to expose the internal force. Second, set up the topology optimization model with the minimum structural quality as the goal and the internal force as the constraint, based on the independent, continuous and mapping method. Then, make sure the unit load method makes the internal force explicit and accumulates the internal force on the force transmission path that needs to be controlled. Finally, iteratively adjust the internal force on each path so that the ratio reaches a stable value, thereby obtaining a force transmission path that satisfies the internal force constraint. [147] The traditional topology optimization method cannot control the material distribution in a specific area during the overall optimization process. In response to this problem, Shu Lei et al. [148] proposed a composite domain topology optimization method, that is, to specify different numbers of materials in different areas to meet the needs of automobiles or equipment in different working environments. To solve the problem of boundary coordination and coupling between the numerical sub-structure and the test sub-structure and the dependence between the sub-structures, unique finite element software can be selected for analysis or experiments according to the characteristics of each sub-structure. Haoran et al. [149] proposed a coordination method for interface unit substructures. By introducing boundary forces to the subdomains and establishing balance relations and displacement coordination relations on the boundary, the flexibility of coupling multiple substructures of the interface unit was solved without the need for substructures. The nodes on the structure boundary correspond in unison. A set of virtual units is defined (the set has a limited number of virtual nodes), and these virtual nodes may not correspond to actual nodes. [150] Wang Bo et al. [151] proposed a process for solving the inherent characteristics of the electric spindle based on the impedance coupling substructure method based on the basic principles of the impedance coupling substructure method. To solve the problem of a longer time by direct analysis method of precise sensitivity of large structures, Zhang Bao et al. [152] arranged the node displacements related to the design variables to the back of the total displacement array, reassembled the stiffness matrix, and then performed regional analysis. It blocked and condensed to obtain a sub-structure matrix, thereby significantly improving efficiency. The substructure method can transform high-order linear equations into low-order linear equations and solve them and it can be condensed, degraded, and solved in stages. It can meet the needs of finite element analysis of large complex structures. [153] Because the number of elements in the finite element model of passenger cars is often relatively large, the topological analysis of it requires a lot of computer time and human resources. Zhang Fan et al. [154] introduced the sub-structure method in the topology optimization analysis of an air-suspension bus. The parts that do not need to be topologically optimized are condensed to generate super-elements through a matrix, and the parts to be optimized are connected with the super-elements to establish a topology model. This greatly reduces the number of model units. In the dynamic structural optimization design, there are often multiple design variables to adjust, but for each variable, the change of its value has a different effect on the structural performance. Choosing the most sensitive variable that affects the dynamic characteristics of the structure as the main parameter for adjustment is of great significance for improving the dynamic characteristics of the structure. However, the huge amount of analysis and long machine time will bring great difficulties to actual operation. To this end, Zhang Zaofa et al. [155] proposed a convenient and effective sub-structure method based on static sensitivity analysis of stress values. To eliminate the system limitation of the dynamic substructure method in solving nonlinear structures, Faye Wang et al. [156] proposed a modal synthesis method suitable for the coupling of nonlinear substructures and multiple linear substructure boundaries. It divides the overall system into two types of linear and nonlinear substructures, and reduces the degree of freedom of the linear substructure according to the potential energy criterion truncation criterion, and then combines it with the nonlinear substructure to obtain the dynamic response of the nonlinear system. Aiming at the problem that it is difficult to determine the boundary conditions of components in a complex mechanical system, which leads to the failure to obtain optimal results, Ding Xiaohong et al. [20] based on the substructure method, determined the load on the components according to the force transmission path and then ensured the component boundary conditions. Under the premise of accuracy, the topology of the component material distribution is finally obtained through gradual approximation and optimization.

From the above literature analysis, it can be seen that the sub-structure method has developed rapidly and has been applied to engineering technology fields such as topology optimization. However, there is no relevant literature on the general research of using the substructure method to optimize large-scale continuous structures. Based on this, this chapter proposes a continuous structural optimization method based on the local feature substructure method. It adopts three types of substructures with different local characteristics to respectively undertake the three types of functions in the optimization process. Each type of substructure will improve the performance of different subfunctions in the optimization process so that the overall optimization has high efficiency and convergence.

5.1 Continuous structural optimization problem description

The research object of this chapter is the structure mass minimization problem of continuous structures (such as a diesel piston) under the constraint of allowable stress and allowable displacement. The optimization model for this is as follows:

$$\begin{cases} \min W(\boldsymbol{X}) \\ \text{s.t.} |\sigma|_{\max}(\boldsymbol{S},\boldsymbol{X}) \leqslant \sigma_{\mathrm{u}} \\ \text{s.t.} |\boldsymbol{d}|_{\max}(\boldsymbol{S},\boldsymbol{X}) \leqslant d_{\mathrm{u}} \\ \text{s.t.}\ \boldsymbol{K}(\boldsymbol{X})\boldsymbol{U}=\boldsymbol{F} \end{cases} \tag{5.1}$$

Where W is the overall quality of the structure; $\boldsymbol{X}$ is the design variable vector; $\boldsymbol{S}$ is the state variable vector; $|\sigma|_{\max}(\boldsymbol{S}, \boldsymbol{X})$ is the absolute maximum stress of the structure; σ_{u} is the upper limit of allowable stress; $|\boldsymbol{d}|_{\max}(\boldsymbol{S}, \boldsymbol{X})$ is the absolute maximum displacement of the structure; d_{u} is the upper limit of allowable displacement; $\boldsymbol{K}(\boldsymbol{X})$ is the total stiffness matrix; $\boldsymbol{U}$ is node displacement vector; and $\boldsymbol{F}$ is power vector.

In continuous structure optimization, the gradient-based optimization method is usually adopted, and the gradient calculation is generally adopted by the central difference method. This method can calculate the gradient for $2n$ times of simulation (n is the number of design variables), and this structural simulation takes a long time. Therefore, the efficiency can be improved from the perspective of structure simulation.

The substructure method is an effective method to improve the efficiency of structure's simulation.

5.2 Substructure method

The substructure method is the process of condensing a group of elements into a super-element with a matrix. A supercell is used in the same way as any other cell. The substructure method can save a lot of computing time and solve large-scale problems with limited computer resources, to improve efficiency. It shows great advantages in nonlinear analysis and optimization analysis of various dynamic analyses with repeated iteration. The substructure method can be divided into two types: fixed interface mode synthesis method and free interface mode synthesis method. As shown in Figure 5-1, the whole structure is divided into two sub-structures I_1 and I_2 along the dotted line, then the dotted line B constitutes the boundary adjacent to each sub-structure and other sub-structures.

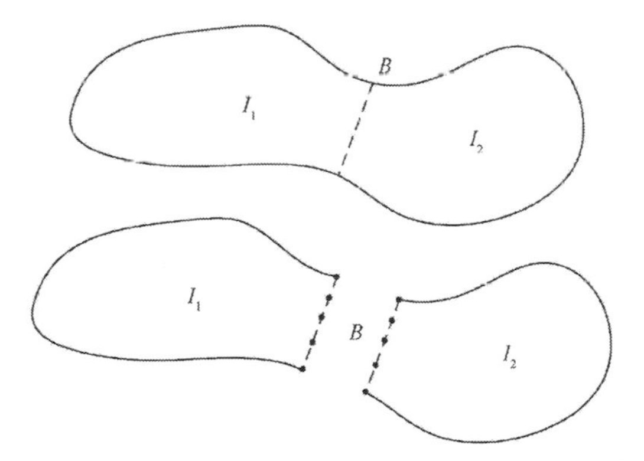

Figure 5-1 Diagram of sub-structure division

The static analysis of the general structure of the global finite element model equation is $K(X)U = F$ in Equation (5.1). First, it is decomposed into two parts: the substructure and the interface between the substructures:

$$\begin{bmatrix} \boldsymbol{K}_{\text{II}}^{(1)} & \boldsymbol{K}_{\text{IB}}^{(1)} & \boldsymbol{0} \\ \boldsymbol{K}_{\text{BI}}^{(1)} & \boldsymbol{K}_{\text{BB}}^{(1)} + \boldsymbol{K}_{\text{BB}}^{(2)} & \boldsymbol{K}_{\text{IB}}^{(2)} \\ \boldsymbol{0} & \boldsymbol{K}_{\text{BI}}^{(2)} & \boldsymbol{K}_{\text{II}}^{(2)} \end{bmatrix} \begin{Bmatrix} \boldsymbol{U}_{\text{I}}^{(1)} \\ \boldsymbol{U}_{\text{B}} \\ \boldsymbol{U}_{\text{I}}^{(2)} \end{Bmatrix} = \begin{Bmatrix} \boldsymbol{F}_{\text{I}}^{(1)} \\ \boldsymbol{F}_{\text{B}} \\ \boldsymbol{F}_{\text{I}}^{(2)} \end{Bmatrix} \tag{5.2}$$

In this equation, $K_{II}^{(i)}$ is the stiffness matrix formed by the internal nodes of the ith sub-structure; $K_{BB}^{(i)}$ is the stiffness matrix composed of interface nodes of the ith substructure; $K_{IB}^{(i)}$, $K_{BI}^{(i)}$ is the connection matrix composed of the internal nodes of the ith substructure and interface nodes; $U_{I}^{(i)}$ is the displacement vector of the internal node of the i th sub-structure; U_B is the displacement vector of the interface nodes of the substructure; $F_{I}^{(i)}$ is the load vector of the internal node of the i th sub-structure; F_B is the load vector of the interface nodes of the substructure. Expand equation (5.2) to get:

$$\begin{cases} \boldsymbol{K}_{II}^{(1)}\boldsymbol{U}_{I}^{(1)} + \boldsymbol{K}_{IB}^{(1)}\boldsymbol{U}_{B} = \boldsymbol{F}_{I}^{(1)} \\ \boldsymbol{K}_{BI}^{(1)}\boldsymbol{U}_{I}^{(1)} + \left(\boldsymbol{K}_{BB}^{(1)} + \boldsymbol{K}_{BB}^{(2)}\right)\boldsymbol{U}_{B} + \boldsymbol{K}_{IB}^{(2)}\boldsymbol{U}_{B} = \boldsymbol{F}_{B} \\ \boldsymbol{K}_{BI}^{(2)}\boldsymbol{U}_{B} + \boldsymbol{K}_{II}^{(2)}\boldsymbol{U}_{I}^{(2)} = \boldsymbol{F}_{I}^{(2)} \end{cases} \tag{5.3}$$

Then the internal nodes of the substructure are reduced, and the substructure reduction equation is obtained as follows:

$$\tilde{\boldsymbol{K}}\boldsymbol{U}_{B} = \tilde{\boldsymbol{F}} \tag{5.4}$$

Where $\tilde{\boldsymbol{K}}$ is the reduced stiffness matrix and $\tilde{\boldsymbol{F}}$ is the reduced load vector. They can be expressed separately as:

$$\tilde{\boldsymbol{K}} = \left(\boldsymbol{K}_{BB}^{(1)} + \boldsymbol{K}_{BB}^{(2)}\right) - \boldsymbol{K}_{BI}^{(1)}\left(\boldsymbol{K}_{II}^{(1)}\right)^{-1}\boldsymbol{K}_{IB}^{(1)} - \boldsymbol{K}_{IB}^{(2)}\left(\boldsymbol{K}_{II}^{(2)}\right)^{-1}\boldsymbol{K}_{BI}^{(2)} \tag{5.5}$$

$$\tilde{\boldsymbol{F}} = \boldsymbol{F}_{B} - \boldsymbol{K}_{BI}^{(1)}\left(\boldsymbol{K}_{II}^{(1)}\right)^{-1}\boldsymbol{F}_{I}^{(1)} - \boldsymbol{K}_{IB}^{(2)}\left(\boldsymbol{K}_{II}^{(2)}\right)^{-1}\boldsymbol{F}_{I}^{(2)} \tag{5.6}$$

The displacement of the interface nodes of the substructure can be obtained by solving Equation (5.4), and then substituted into the corresponding part of Equation (5.2), and the displacement of the internal nodes of each substructure can be obtained.

From the above analysis, it can be seen that the order of the matrix $\tilde{\boldsymbol{K}}$ is much smaller than that of matrix K, so the scale of the solution formula (5.4) is much smaller than that of the solution formula (5.1), which is the appeal of the substructure method.

5.3 Implementation of the substructure method

The optimization process always requires iterative calculation, and the substructure method can greatly reduce the calculation scale. If the two are combined, unexpected results will be obtained. Combined with the characteristics of optimization, the analysis process of the substructure method is described as follows.

1) Determine the substructure. Divide the continuous structure to be optimized into two categories of substructures: The first is a substructure that contains neither state variables nor design variables, called the super-element. The second is a substructure that contains state variables and/or contains design variables, called non-super-element. The analysis of the former requires the use of a substructure method, while the latter does not.

   In the process of optimization, the geometric structure of the first substructure does not change, and the second substructure can be divided into two types: a) Parametric substructure; b) State variable substructure. Within a structure, there can be multiple parametric substructures and multiple state variable substructures. Parametric substructures are local structures that contain design variables. State variable substructure refers to the local structure containing state variables, that is, its constraint function or objective function contains state variables, but does not contain design variables. Sometimes parameterized substructures contain not only design variables but also state variables, but state variable substructures contain only state variables.

   As shown in Figure 5-2, the overall structure is divided into three sub-structures $I_i = (i = 1, 2, 3)$, and B is the boundary adjacent to each sub-structure and other sub-structures. The nodes on each sub-structure are connection nodes and the rest are internal nodes. The I_1 is a substructure of the first class, which geometric structure does not change during optimization and contains neither design variables nor state variables. The I_2, I_3 is the second substructure, where I_2 is a parametric substructure and I_3 is a state variable substructure.

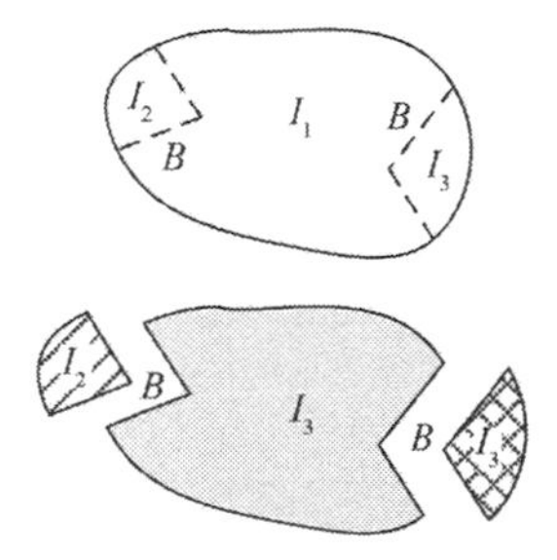

Figure 5-2 Schematic diagram of optimized substructure

2) Selective principal degree of freedom. The master degrees of freedom include: a) All nodes of the part in contact with a non-supercell; b) All nodes corresponding to the constraint condition; c) All nodes where the load is applied; d) Arbitrary nodes (the

selection of which is related to precision). For general structures, only a)b)c) can be chosen as the main degree of freedom, but in some special cases, in addition to a)b)c) as the main degree of freedom, the main degree of freedom related to d) should also be considered.

3) The different substructures are generated into supercells. According to Equations (5.2–5.5), matrix reduction is carried out for each substructure, so that the combined overall matrix scale is greatly reduced.
4) Coupling super-element and non-super-element. During the optimization process, the geometric structure of the supercell does not change. Before optimization, mesh division and boundary conditions are applied to this molecular structure first, so that the optimization can be used directly, which will save a lot of time. The non-super-element can be divided into two types. The first is parametric substructure, and the boundary conditions of its meshing should be applied in real-time. The most important part of this molecular structure is the connecting part, the common part, between the super-element and the non-super-element. In coupling, the nodes corresponding to the common part of the supercell should be divided as initial conditions for the non-supercell, and then the common nodes corresponding to the two structures should be coupled after the division. The other is the substructure of the state variable, which is treated like that of the super-element, that is, its geometric structure does not change during optimization. Therefore, before optimization, this molecular structure is first meshed and applied with boundary conditions, and then coupled with the supercell, which can be directly applied in the optimization process, thus saving a lot of time. In this way, it can avoid the need to extend the super unit after each simulation, to obtain the required value of the state variable and greatly improve the computational efficiency.
5) Solving. The global matrix size of the coupled structure is much smaller than that of the non-substructure. In general, in a continuous structure, the proportion of super-elements is much larger than that of state variable substructures and parametric substructures. Therefore, the application of substructures can greatly reduce the computational scale and improve computational efficiency.
6) Extension. This extension is time-consuming and can be done at the end when dealing with problems that need to be solved repeatedly. For example, in the optimization process, the structural finite element model needs to be solved repeatedly, which can be verified and analyzed by extension after the optimization. The purpose of applying state variable substructure is to avoid the computation of substructure extension during optimization, thus saving a lot of time and improving efficiency.

5.4 Optimization iteration based on substructure method

The optimization iteration process based on the substructure method is shown in Figure 5-3. When dividing a substructure, if the state variable is not in a separate substructure but also included in a parameterized substructure, then the state variable substructure is not constructed. The optimization iteration process, in this case, shown in Figure 5-3 (a). Figure 5-3 (b) is a general optimization iterative process, including three parts: super-element, state variable substructure, and parametric substructure. The finite element model corresponding to the first two parts only needs to be generated once, and it is coupled with the third part in the iterative process to participate in the global model solution. After the solution is completed, the value of the state variable shall be extracted from the state-variable quantum structure, and the value of the state variable shall be extracted from the parametric sub-structure, and then all the value of the state variable and the initial value of the design variable of the parametric sub-structure shall be returned to the optimizer. It will analyze and calculate and determine the value of the design variable of the next iteration. In the optimization process, the parametric sub-structure needs to generate a geometric structure in real-time, divide the grid, impose boundary conditions, and couple the contact boundary with other sub-structures. In other words, the parametric sub-structure is reconstructed by using the newly designed variable value and iteratively iterates until convergence.

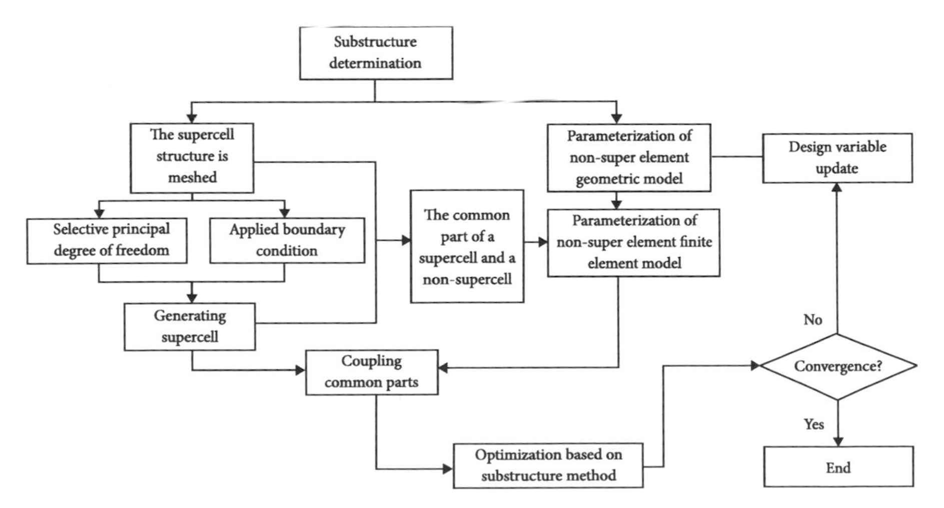

Figure 5-3 Optimization iteration process based on sub-structure method

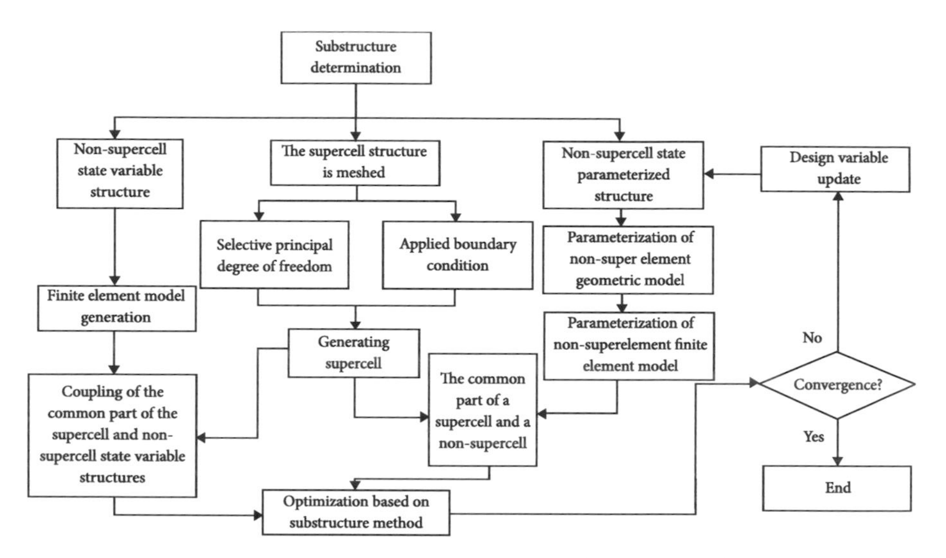

Figure 5-3 Optimization iteration process based on sub-structure method (continued)

This step of expansion cannot be found in Figure 5-3, because the substructure has been fully considered from the perspective of optimization, and the overall structure has been divided into three categories: super-element, state variable substructure, and parameterized substructure. In the optimization process, the values of the state variables required by the objective function and the constraint function are only obtained from the state variable sub-structure and the parameterized sub-structure, so there is no need to expand the super-unit. Expansion is time-consuming, and super unit expansion after optimization is done to verify the results of optimization.

5.5 Example analysis

5.5.1 Hollow beam design

There is a hollow beam fixed at both ends, and its structure is shown in Figure 5-4. Its thickness is 0.1m, the material's elastic modulus is, and the Poisson's ratio is 0.3. The center of the upper part of the hollow web is subjected to a vertical downward concentrated load, and its size is 1000N. We need to determine the size of the hollow web beam so that its mass is minimal under the conditions of displacement and stress constraints.

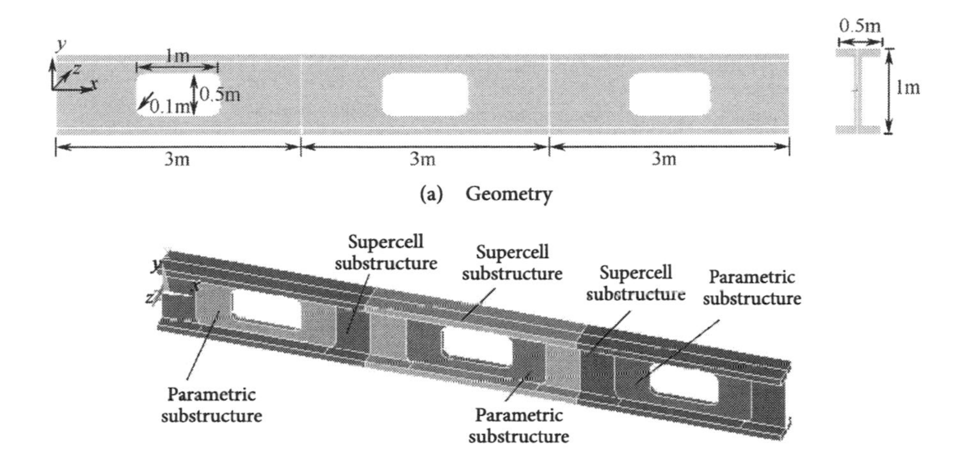

(a) Geometry

(b) Substructure division

Figure 5-4 Hollow beam structure

The optimization model is shown in Equation (5.7)

$$
\begin{cases}
\min W \\
\text{s.t.} \dfrac{|\upsilon|_{\max}}{4.6\times10^{5}}-1\leqslant 0 \\
\text{s.t.} \quad \dfrac{|\boldsymbol{d}|_{\max}}{0.000026}-1\leqslant 0 \\
\text{s.t.} \quad \boldsymbol{KU}=\boldsymbol{F}
\end{cases}
\tag{5.7}
$$

Figure 5-5 shows the iterative process of optimization of the hollow beam based on the substructure method. Among them, Figure 5-5 (a) is the iterative process of the objective function, and Figure 5-5 (b) is the iterative process of the constrained function of the response. It can be seen from Figure 5-5 that the changing trend of the mass of the empty web corresponds to the default ratio. If the default ratio is maintained or decreased, the mass of the fasting beam will also be maintained or decreased. Once the default ratio is increased, the mass of the empty beam will also increase significantly. Figure 5-6 shows the iterative process of direct optimization of the open web beam. From Figure 5-6 (a), the mass of the fasting beam seems to decrease. As can be seen from Figure 5-6 (b), the default is relatively large, reaching 2.69e-02. Figure 5-7 is the verification of the optimization results based on the substructure method. It can be seen from the figure that it meets the constraints. From Table 5.1 to Table 5.4, it can be seen that the average simulation time of optimization based on the substructure method is less than that of direct optimization. The average time of the former is 12.5 s, while that of the latter is 14.35 s. The substructure method has no obvious advantage because the structure is smaller and more regular.

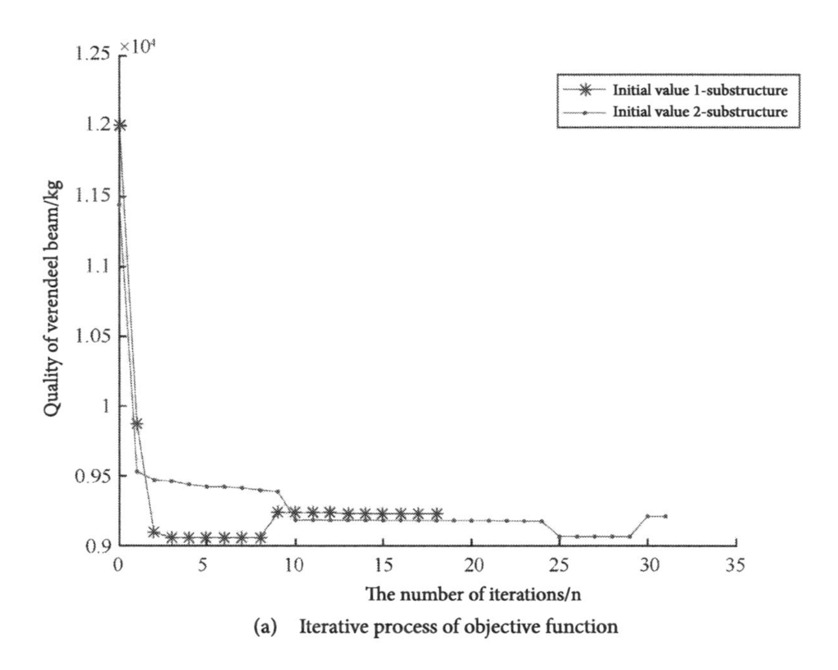

(a) Iterative process of objective function

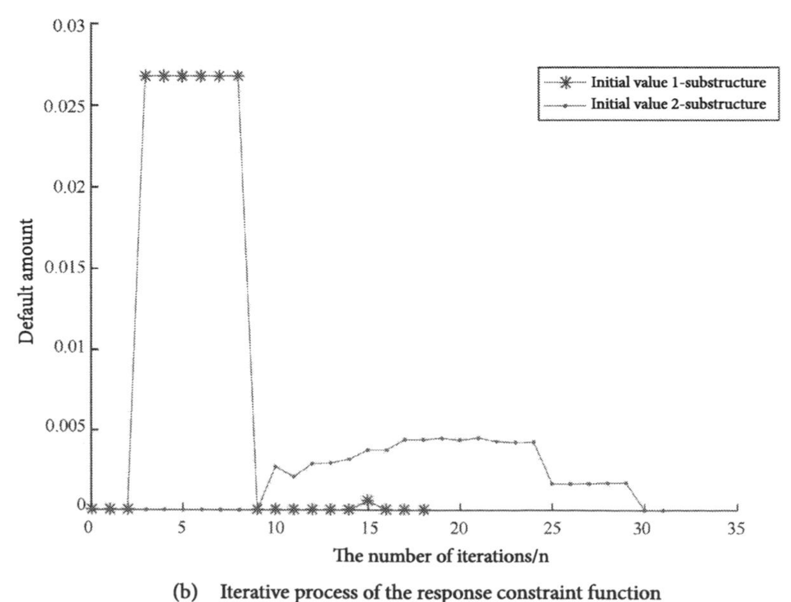

(b) Iterative process of the response constraint function

Figure 5-5 Iterative process of hollow beam optimization based on sub-structure method

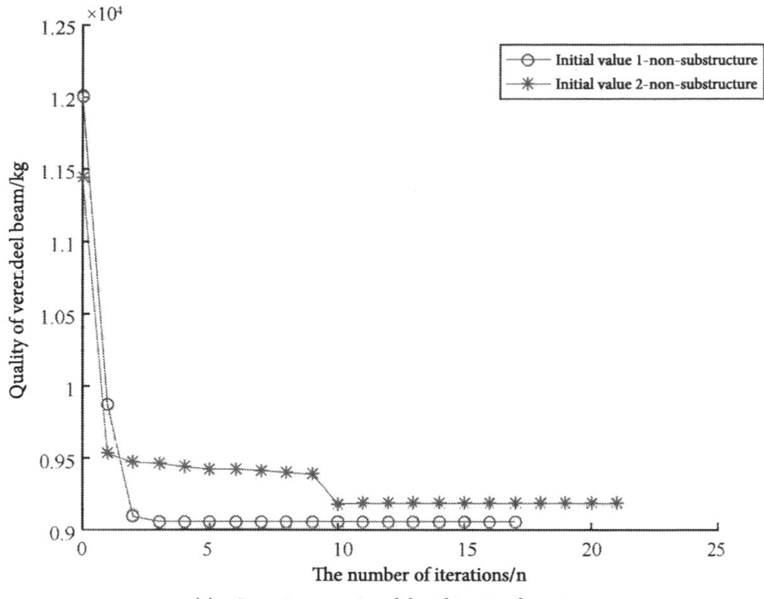

(a) Iterative process of the objective function

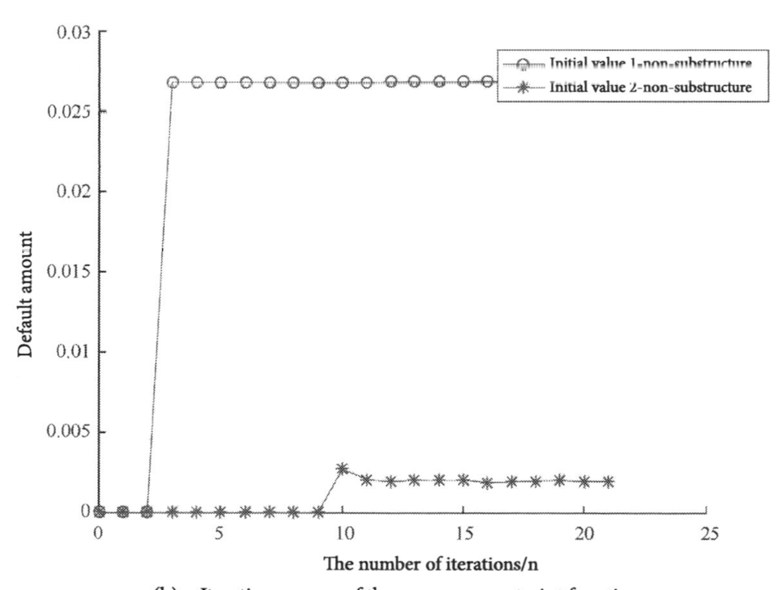

(b) Iterative process of the response constraint function

Figure 5-6 Iterative process of direct optimization of hollow beam

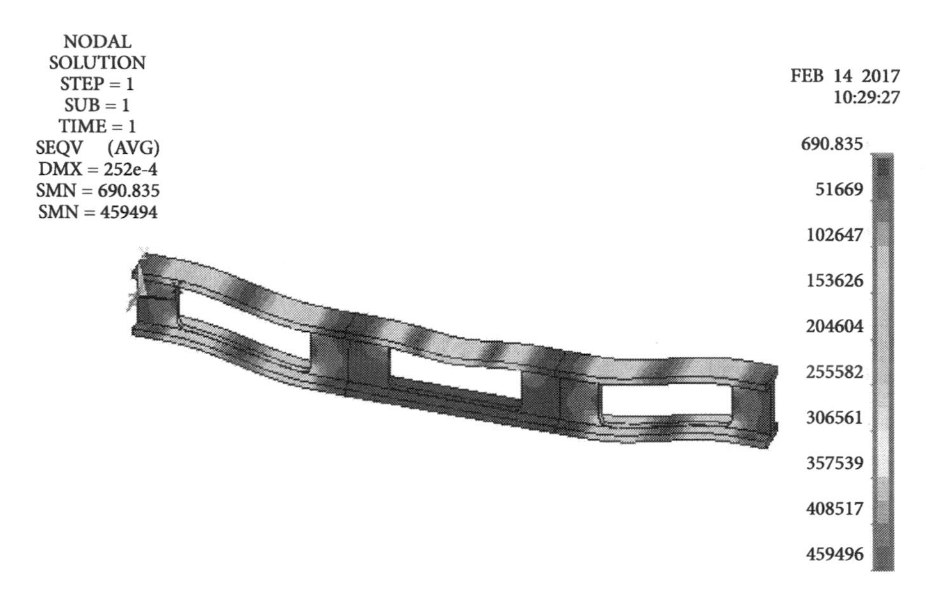

Figure 5-7 Verification of the optimization results based on the sub-structure method (in the case of initial value 2)

Table 5.1 Optimal values of hollow beams optimized based on sub-structure method

Project		The design variables		
		x_1	x_2	x_3
The initial value	1	0.80000000	0.40000000	0.15000000
	2	1.40000000	0.40000000	0.15000000
Optimal value	1	1.71864084	0.89943379	0.15420764
	2	1.89997288	0.79025765	0.11905143

Table 5.2 Iteration of hollow beam optimization based on sub-structure method

Project	The number of iterations/time	The simulation number of times/time	Objective function value /kg	Constrained function value	The total time/s	Mean simulation time/s
1	18	108	9230.45	0.0	1346	12
2	31	148	9211.10	0.0	2017	13

Table 5.3 Optimal values for direct optimization of open web beams

Project		The design variables		
		x_1	x_2	x_3
The initial value	1	0.80000000	0.40000000	0.15000000
	2	1.40000000	0.40000000	0.15000000
The optimal value	1	1.89999918	0.89999686	0.14604088
	2	1.89996966	0.79525400	0.11061828

Table 5.4 Iteration of direct optimization

Project	The number of iterations/time	The simulation number of times/time	Objective function value/kg	Constrained function value	The total time/s	Mean simulation time/s
1	17	86	9058.48	2.690e-02	1277	14.8
2	21	116	9184.63	1.962e-03	1619	13.9

The optimization of the hollow beam based on the substructure method belongs to the optimization of substructures that do not contain non-super-element state variables because the state variables required by the constraint function are included in the parameterized substructure, so there is no need to construct state variables in the optimization process. Substructures obtain the value of the required state variable, which can save processing time and further reduce the scale of solving equations.

5.5.2 Diesel engine piston design

The geometric structure of a diesel engine piston is shown in Figure 5-8. The elastic modulus of the main material is 1, and Poisson's ratio is 0.3. The elastic modulus of the ring material is 2, and Poisson's ratio is 0.33. To explain the superiority of the substructure method, the following uses the minimum piston mass as the objective function and the oil cavity shape as the design variable for optimization.

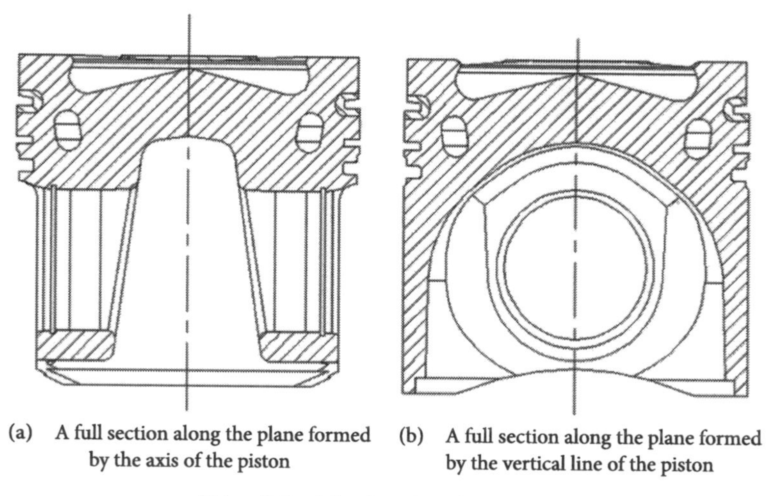

(a) A full section along the plane formed by the axis of the piston (b) A full section along the plane formed by the vertical line of the piston

Figure 5-8 Diesel engine piston geometry

This book does not consider the cylinder when modeling, so ignore the influence of lateral thrust and friction on the piston. Mechanical stress mainly includes the effect of maximum burst pressure and piston inertial force. The boundary conditions of the gas pressure load of the piston are shown in Figure 5-9. Combining the actual force of the piston, the pressure P_j in the cylinder is applied to the top surface of the piston according to the uniform distribution force and it is applied to each ring groove and ring bank in a certain proportion. The inertial force of the reciprocating piston can be calculated according to the acceleration curve of the piston. Because the full constraint on both ends of the piston pin is calculated by using a 1/4 finite element model, the symmetric constraint is applied to the symmetric plane. In the static mechanical analysis, considering the force of the piston at the moment of maximum explosion pressure, and according to the stress distribution in the previous trial calculation, the piston is divided into three substructures using the substructure method, namely the super-element, the state variable substructure and the parameterized substructure, and then the mesh, as shown in Figure 5-10.

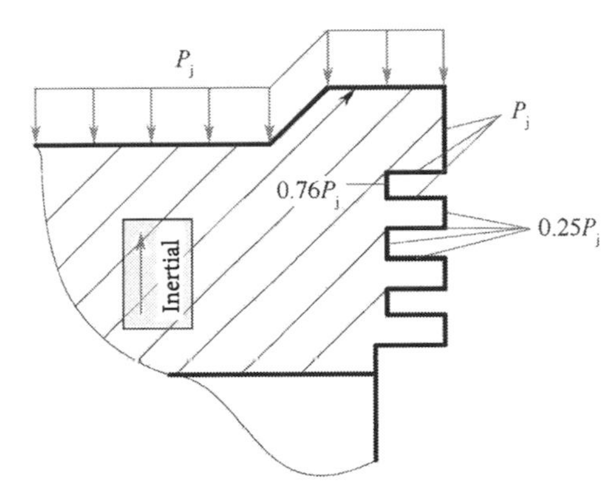

Figure 5-9 Boundary conditions of piston gas pressure load

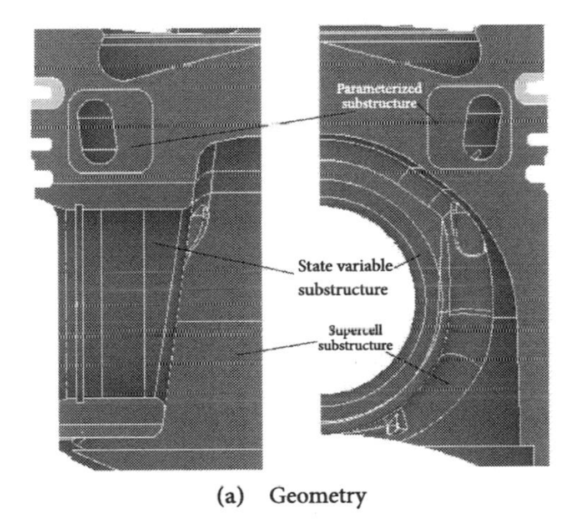

(a) Geometry

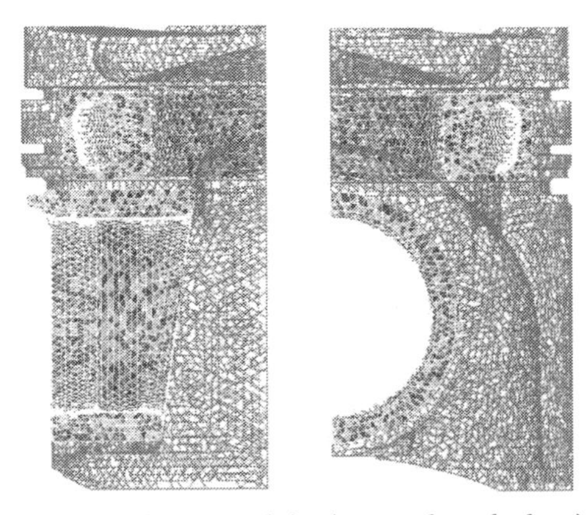

(b) Finite element mesh (in the case of initial value 5)

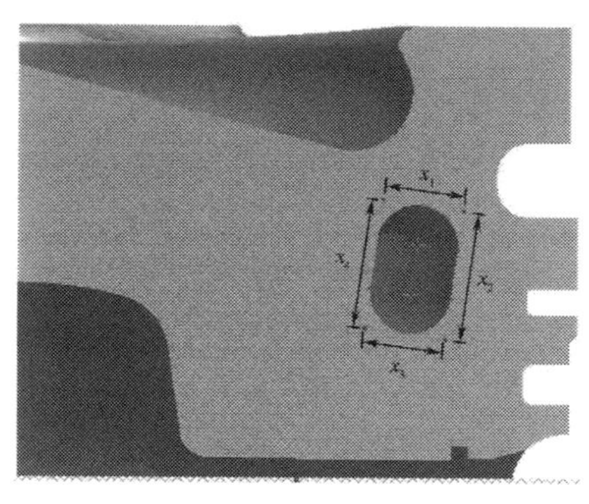

(c) Design variable

Figure 5-10 Diesel engine sub-structure division and design variables (continued)

The finite element simulation calculation is based on the three-dimensional solid model of the structure. On the premise of ensuring the accuracy of the simulation calculation, to reduce the calculation time, this chapter uses the 1/4 piston combination model as the finite element calculation model and uses the Solid45 unit for meshing. The finite element mesh model used for the initial piston calculation is shown in Figure 5-10 (b). The model contains 123011 tetrahedral elements and 24695 nodes. The optimization model of the diesel engine piston is shown in equation (5.8).

$$\begin{cases} \min W \\ \text{s.t.}\ \dfrac{|\sigma|_{\max}}{190} - 1 \leqslant 0 \\ \text{s.t.}\ \dfrac{|\boldsymbol{d}|_{\max}}{0.365} - 1 \leqslant 0 \\ \text{s.t.}\ \boldsymbol{KU} = \boldsymbol{F} \end{cases} \tag{5.8}$$

In the optimization process of the diesel engine piston based on the substructure method, the fmincon function in MATLAB optimization toolbox is used as the optimizer and combined with the finite element solver for optimization. For the diesel engine piston, after preliminary calculation, the state variables required by the constraint function are not included in the parameterized substructure. Therefore, the state variable substructure needs to be constructed independently, as shown in Figure 5-10 (a). There are four design variables for the diesel engine piston, as shown in Figure 5-10 (c).

To illustrate the effectiveness of the optimization method based on the substructure method, and to exclude the influence of the initial value on the results, 5 points are taken arbitrarily in the design domain as the initial value for optimization. The selection of initial values and optimization results are shown in Table 5.5. The optimization iteration process corresponding to the 5 sets of initial values is shown in Figure 5-11.

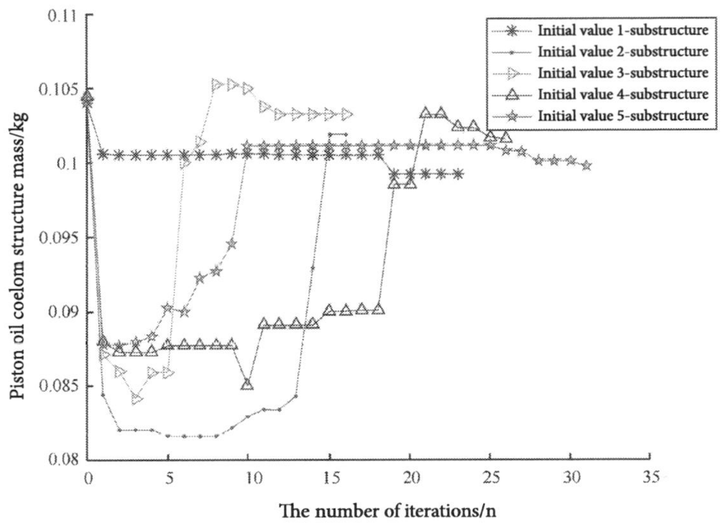

(a) Iterative process of the objective function

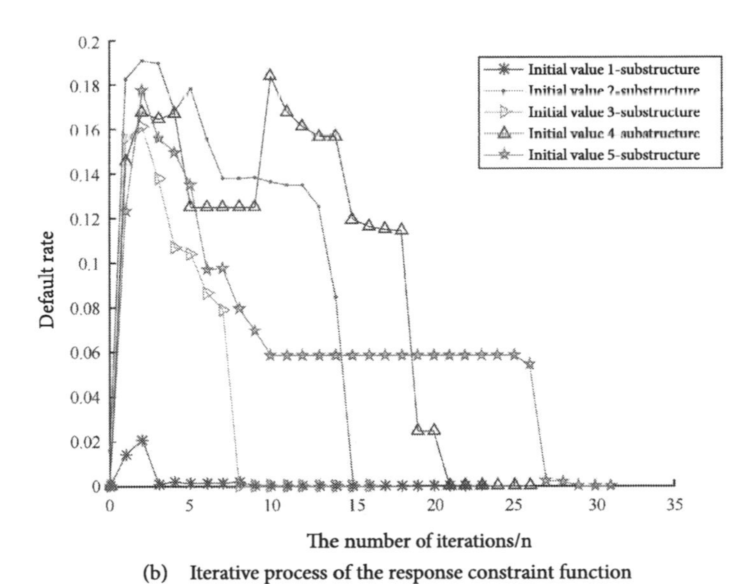

(b) Iterative process of the response constraint function

Figure 5-11 Iterative process of diesel engine piston optimization based on sub-structure method

It can be seen from Figure 5-11, Figure 5-12, and Table 5.5–Table 5.8 that the initial value is different and the optimal value is also different, which is caused by local optimization. Moreover, the function composed of the design domain composed of the objective function and the constraint function is a black-box function, since we do not know its characteristics, but it can be safely said that it also serves a multi-peak function. To obtain the global optimal solution, a global optimization solver is required. However, currently, the global optimization solver such as genetic algorithm, simulated annealing algorithm, and particle swarm algorithm has a large number of simulations, and it is directly applied to solve engineering problems. Its success rate is very small. The simulation times of the gradient-based optimization solver and modern optimization solver are acceptable, but the results are locally convergent. Therefore, only by selecting multiple initial points can the global optimal solution be obtained as much as possible.

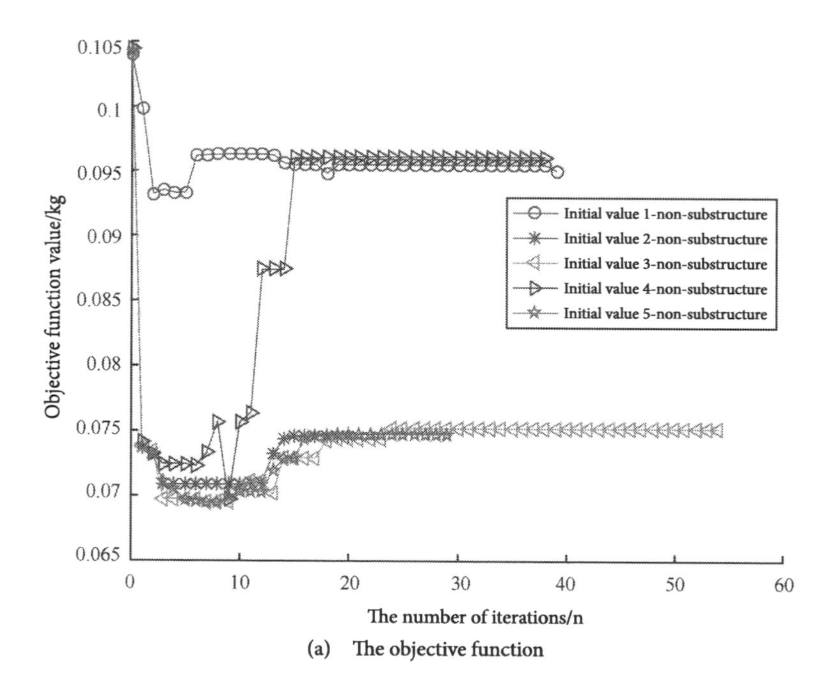

(a) The objective function

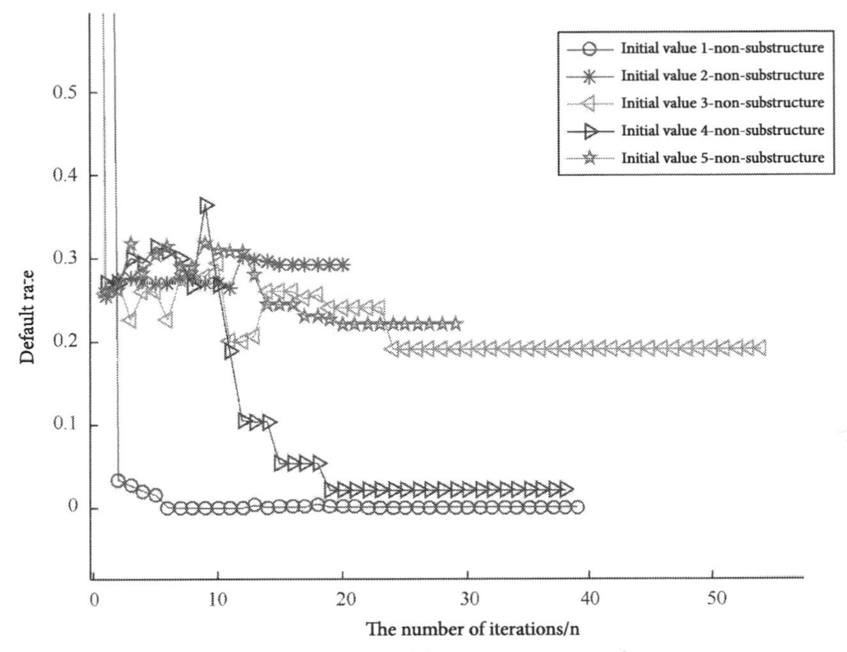

(b) The iterative process of the response constraint function

Figure 5-12 Iterative process of direct optimization (continued)

Table 5.5 Optimal values of diesel engine pistons optimized based on sub-structure method

Project		The design variables			
		x_1	x_2	x_3	x_4
The initial value	1	−0.20000000	−0.80000000	1.80000000	5.90000000
	2	−0.40000000	−0.80000000	1.80000000	5.90000000
	3	−0.60000000	−0.80000000	1.80000000	5.90000000
	4	−0.90000000	−0.80000000	1.80000000	5.90000000
	5	−1.20000000	−0.80000000	1.80000000	5.90000000
The optimal value	1	−0.00000021	−0.00341960	1.72226711	4.68397495
	2	−1.19999973	−0.59552277	0.00131510	4.38518460
	3	−0.81972655	−0.79999994	0.00595804	4.37905906
	4	−0.00475140	−0.59773221	0.87076402	4.37696719
	5	−0.01920395	−0.41019366	1.01309146	4.31331118

Table 5.6 Iteration of diesel engine piston optimization based on sub-structure method

Project	Number of iterations/ time	Simulation times/time	Objective function value/kg	Constrained function value	The total time/s	Mean simulation time/s
1	23	225	0.0991920	0.0	54536	242
2	16	124	0.1018669	0.0	35230	284
3	16	151	0.1032603	0.0	38377	254
4	26	220	0.1015546	0.0	46698	212
5	31	240	0.0996922	0.0	44701	186

Table 5.7 The optimal value of direct optimization of the diesel engine piston

Project		Design variable			
		x_1	x_2	x_3	x_4
Initial value	1	−0.20000000	−0.80000000	1.80000000	5.90000000
	2	−0.40000000	−0.80000000	1.80000000	5.90000000
	3	−0.60000000	−0.80000000	1.80000000	5.90000000
	4	−0.90000000	−0.80000000	1.80000000	5.90000000
	5	−1.20000000	−0.80000000	1.80000000	5.90000000
Optimization value	1	−0.00005292	−0.19746672	0.37001164	4.00974724
	2	−0.06380442	−0.68293880	0.00000000	0.00000003
	3	−0.26755315	−0.67042140	0.01867824	0.09127530
	4	−0.00000002	−0.79999999	0.00000010	3.36922375
	5	−0.41698987	−0.66398598	0.00002144	0.00000000

Table 5.8 Iteration of direct optimization of diesel engine piston

Project	Number of iterations/ times	Simulation times/times	Objective function value/kg	Constrained function value	The total time/s	Mean simulation time/s
1	39	422	0.0950418	1.015e-03	167013	395
2	20	135	0.0746029	2.931e-01	76472	566
3	54	417	0.0752296	1.908e-01	166921	400

(Continued)

Project	Number of iterations/ times	Simulation times/times	Objective function value/kg	Constrained function value	The total time/s	Mean simulation time/s
4	38	259	0.0961158	2.115e-02	107301	414
5	29	241	0.0747647	2.209e-01	106182	440

Figure 5-11 is the iterative process of diesel engine piston optimization based on the sub-structure method. As can be seen from Figure 5-11 (a), the optimization process fluctuates greatly, which corresponds to the fluctuation of Figure 5-11 (b). It can be seen from Table 5.6 that the optimization based on the substructure method finally converges into the feasible region, that is, all satisfy the constraint function. Its optimal objective function value is 0.0991920 kg (only when referring to the part of parametric sub-structure). When reaching this value, 23 optimization iterations and 225 simulation times have been achieved, which takes 54536 s and 242 s on average for one simulation. Among the 5 optimizations with different initial values, the average time used most is 284 s/time. Figure 5-12 is the iterative process of piston direct optimization. It can be seen from Figure 5-12 (a) that the fluctuation of the optimization process is not as great as that in Figure 5-11 (a), which is of course also corresponding to Figure 5-12 (b). However, as can be seen from Table 5.8, none of its optimal solutions satisfy the constraint function, that is, the constraint function values are not equal to zero, and the minimum value is 1.015e-03. The minimum objective function value is 0.0746029 kg, and the corresponding constraint function value is 2.931e-01. At the same time, direct optimization with the same optimization conditions sees the overall average of 5 optimizations at 443 s per simulation. Judging from the average time spent per simulation, the optimization method based on the substructure method takes 53.18% of the direct optimization method, which shows that it saves 46.82% of the time and greatly improves efficiency.

Figure 5-13 shows the verification of the optimization results of the diesel engine piston based on the substructure method.

In this section, the optimization of the diesel engine piston is only used to illustrate the advantages of optimization based on the substructure method, so the simulation part is partially simplified. This means that the dynamic optimization is simplified to static optimization, and the temperature load is not considered.

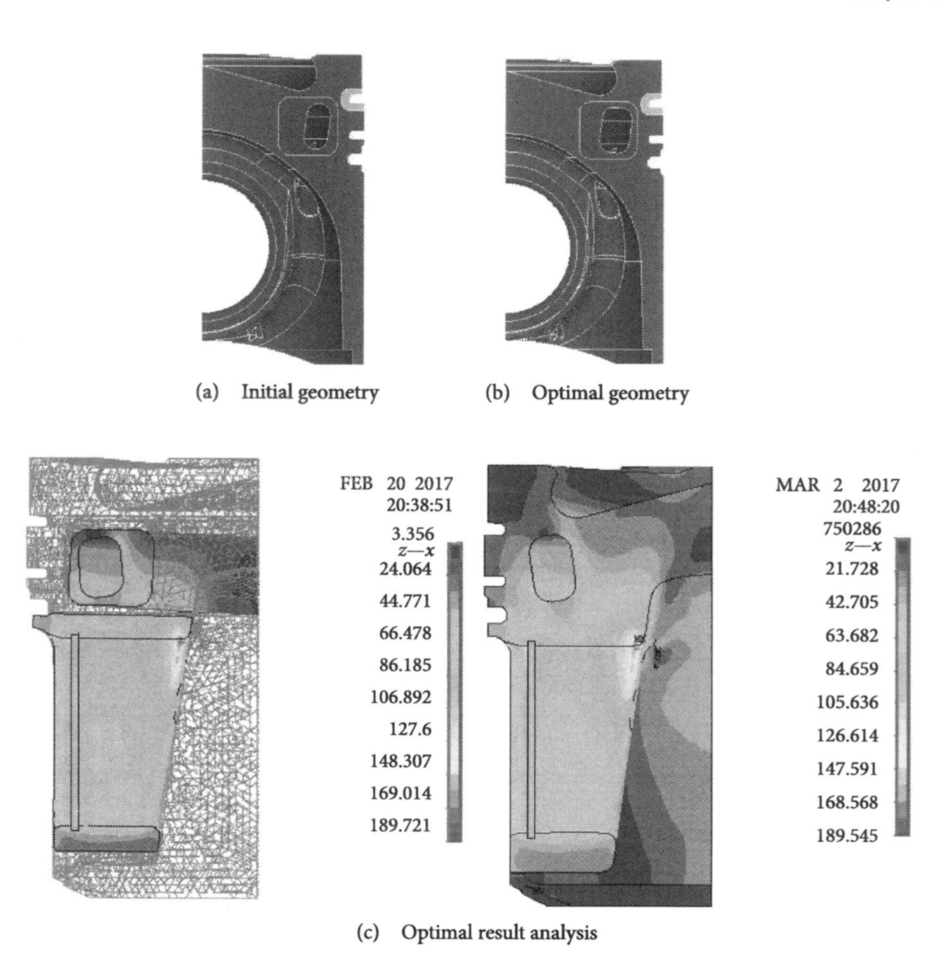

(a) Initial geometry (b) Optimal geometry

(c) Optimal result analysis

Figure 5-13 Verification of the optimization results of diesel engine pistons based on the sub-structure method (the initial value of 5)

5.6 Summary of this chapter

This chapter divides the continuous structure into super-units, state variable sub-structures, and parameterized sub-structures according to the needs of optimization, and then reduces the super-units to be fully coupled with the state variable sub-structure or with the parameterized sub-structure, which greatly reduces the calculation scale. In addition, this chapter illustrates the optimization method based on the substructure method which is very effective through two examples of optimization of the hollow beam and the optimization of the piston of the diesel engine. The following conclusions are obtained:

1) The local feature sub-structure is a sub-structure that divides different parts of the continuous structure according to the main tasks it undertakes in the optimization process. The main task of the super-element is to reduce the simulation calculation scale, while the main task of the state variable substructure is to provide the value of the state variable required by the objective function and the constraint function. Similarly, the main task of the parameterized substructure is to reduce the quality of the structure, which is the ultimate goal of optimization.
2) Continuous structures can be divided into super-units, state variable sub-structures, and parameterized sub-structures according to local characteristics. Super-units account for most of the continuous structures, while state variable sub-structures and parameterized sub-structures account for only a small part. When the structure reaches a certain scale, the time spent on its pre-treatment is almost negligible.
3) Under the coupling effect of the three local feature substructures, the optimization method based on the substructure method not only greatly improves the efficiency, but also converges it as well. For the optimization problem of a diesel engine piston (including 123011 tetrahedral elements and 24695 nodes), this method can save 46.82% of the time and greatly improve the optimization efficiency.

CHAPTER 6

Structural Dynamic Characteristics Optimization Based on Average Element Energy of the Substructure

The primary task of structural optimization is to establish a reasonable optimization model, which includes determining design variables and their value ranges. The value of the design variable directly determines the convergence and efficiency of the optimization problem. There is relatively little literature available for reference in this respect. Engineers usually take more design variables based on experience and expand the value range of design variables as much as possible to avoid the omission of key design variables and their important value ranges. If there are too many design variables, you can delete some design variables that are not sensitive to the target through sensitivity analysis. However, the efficiency of sensitivity analysis for large structures is very low and is not suitable for engineering applications.

The optimization of the dynamic characteristics of the structure, especially the flexible structure of aerospace composite materials [157] is to improve the natural frequency of the main vibration mode and reduce the quality of the structure. In engineering, it is often required to increase the natural frequency of the main vibration mode without changing the volume. Zhao Ning et al. [158] of Northwestern Polytechnical University combined structural dynamic characteristics with shape optimization and proposed a new method for the dynamic shape optimization design of a blade-disk type structure, which satisfies the dynamic conditions such as frequency conditions, stability, and stiffness. Under the premise, the economic performance, process performance, and use performance of the structure are optimized. References [159] establish an optimization model by reasonably selecting design variables, state variables, and objective functions to optimize the dynamic performance of the machine gun structure, making the design frequency of the weapon and the natural frequency of the machine gun's overall structure more reasonable, improving the weapon design's precision. Li Xiaogang et al. [160] comprehensively considered the uncertainty of mechanical structural material property parameters and structural dynamic characteristics, established the mean and variance of mechanical structural dynamic characteristic indexes as optimization objectives, and the mechanical structural dynamic characteristics constrained by the amount of mechanical structural deformation. To robustly optimize the mathematical model, and obtain all the Pareto optimal solutions of the problem through the neighborhood cultivation genetic algorithm and the double-layer updated Kriging model, the literature [161] takes the inclination angle and size of the evenly distributed reinforcement as its design variables. The static characteristics that maximize the critical buckling load and the dynamic characteristics with the largest natural frequency are optimized. The results obtained have important reference values for the vibration and noise control of the evenly distributed rib structure. Literature [162] analyzed the structure dynamic characteristics of the new type of glaze spraying robot, found its main vibration frequency and main vibration mode, and proposed an improvement measure to make the structural rigidity increase and the maximum deformation decrease. References [163] obtain a new optimal solution based on the optimal solution when there is no parameter disturbance, and extrapolate the disturbance by Taylor expansion, and give the design result and the sensitivity curve at the same time, which can improve the credibility of engineering technicians for optimal design to promote the promotion of optimization technology. References [164] use the Rayleigh quotient method to solve the sensitivity problem of the structural dynamic characteristics change caused by the position of the center of gravity of the plate structure and the beam structure. The eigenvalues need to be solved repeatedly during the solution process. References [165] obtained the weak region of the structure through modal analysis and obtained that the strength of the structure can withstand various possible stress conditions based on the Rayleigh quotient method so that the strength of the structure is approximately isotropic. References [166] regard the sum of the squares of the difference between the dynamic characteristic parameters and the target value of the brake to be changed to the objective function, determine the design variables through sensitivity analysis, and determine the variation range of the design variables according to actual conditions and experience. The final optimization result changed the vibration amplitude at the key nodes of the brake bracket substructure and eliminated the system coupling unstable mode.

In summary, the current research on the optimization of structural dynamic characteristics is focused on different solutions for different types of structural optimization problems and it rarely involves the establishment of a more reasonable dynamic characteristics optimization model for general structures to further improve optimized robustness. Therefore, starting from the rationality of the structural dynamic characteristic optimization model, this chapter proposes a structural dynamic characteristic optimization method based on the average element energy of the substructure.

6.1 Description of the structural dynamic characteristics' optimization problem

The research object of this chapter is the minimum volume problem of the structure under the minimum allowable main vibration constraint of the natural frequency mode, or the maximization of the natural frequency of the main vibration mode under the allowable maximum volume constraint, or the constraint of keeping the volume constant. As such, the main natural vibration mode natural frequency maximization problem and the corresponding optimization models are as follows:

$$\begin{cases} \min V(\boldsymbol{X}) \\ \text{s.t. } \det\left(\boldsymbol{K}-\omega^2\boldsymbol{M}\right)=0 \\ \text{s.t. } \omega_1(\boldsymbol{X}) \geqslant \omega_0 \\ \text{s.t. } \boldsymbol{X}_1 \leqslant \boldsymbol{X} \leqslant \boldsymbol{X}_\text{u} \end{cases} \tag{6.1}$$

$$\begin{cases} \max \omega_1 \\ \text{s.t. } V(\boldsymbol{X}) \leqslant V_0 \\ \text{s.t. } \det\left(\boldsymbol{K}-\omega^2\boldsymbol{M}\right)=0 \\ \text{s.t. } \boldsymbol{X}_1 \leqslant \boldsymbol{X} \leqslant \boldsymbol{X}_\text{u} \end{cases} \tag{6.2}$$

$$\begin{cases} \max \omega_1 \\ \text{s.t. } V(\boldsymbol{X}) = V_0 \\ \text{s.t. } \det\left(\boldsymbol{K}-\omega^2\boldsymbol{M}\right)=0 \\ \text{s.t. } \boldsymbol{X}_1 \leqslant \boldsymbol{X} \leqslant \boldsymbol{X}_\text{u} \end{cases} \tag{6.3}$$

Where V is the structure volume; V_0 is the initial volume of the structure; $\boldsymbol{X}$ is the design variable vector; $\boldsymbol{K}$ is the total stiffness matrix; $\boldsymbol{M}$ is the total mass matrix; ω_1 is the natural frequency of the main vibration mode; ω_0 is the minimum main vibration allowed by the structure Modal natural frequency; $\boldsymbol{X}_1$ is the lower limit of the design variable vector; $\boldsymbol{X}_\text{u}$ is the upper limit of the design variable vector.

In structural optimization, usually, when building an optimization model, it is necessary to consider which quantities are design variables and what their value ranges are. For the selection of design variables, the traditional sensitivity analysis method is often used to analyze which parameter changes have a greater impact on the target value, and which parameter changes hardly affect the target value, so that the parameters that have the greatest impact are one of the design variables. The value of the design variables is usually determined based on experience, geometric allowance, physical allowance, and other factors, often taking the largest possible range, but this will bring great challenges to the optimizer and consume a lot of time. As such, determining design variables and their value ranges is very important. This chapter selects design variables and determines their value range from the perspective of the average element energy of the substructure.

6.2 The structure's average element energy

Structural dynamic characteristics optimization takes the natural frequency, vibration mode, or dynamic response of some local points (ranges) of the structure as the objective function or constraint condition, and achieves the expected design goal through the optimization method, and then finally achieves the optimal performance of the structure. Taking the natural frequency of the structure, the vibration mode, or the dynamic response of some local points (range) as the objective function or the constraint function, the essence remains the same, but reduces the dynamic response level of the structure.

6.2.1 Factors affecting the dynamic characteristics of the structure

For the structure subjected to harmonic excitation, the finite element dynamic equation can be expressed as

$$\boldsymbol{M}\ddot{\boldsymbol{X}} + \boldsymbol{C}\dot{\boldsymbol{X}} + \boldsymbol{K}\boldsymbol{X} = \left[00\cdots F_j\cdots 00\right]^{\mathrm{T}} \sin(\omega t) \tag{6.4}$$

In the formula, M is the mass matrix; K is the stiffness matrix; C is the damping matrix; F_j is the external force amplitude of coordinate j; ω is the frequency of the harmonic excitation force. The dynamic response of the structure at point k can be approximated as

$$\boldsymbol{X}_k = \frac{u_{jr}u_{kr}}{\boldsymbol{C}_r}\frac{\sqrt{\boldsymbol{M}_r}}{\sqrt{\boldsymbol{K}_r}}F_j \tag{6.5}$$

In the formula, u_{jr} and u_{kr} are the jth and kth elements of the rth eigenvector respectively; $\boldsymbol{M}_r$ is the modal mass matrix; $\boldsymbol{K}_r$ is the modal stiffness matrix; $\boldsymbol{C}_r$ is the modal damping matrix.

$$M_r = u_r^{\mathrm{T}} M u_{r,}\ K_r = u_r^{\mathrm{T}} K u_{r,}\ C_r = u_r^{\mathrm{T}} C u_r \tag{6.6}$$

Equation (6.5) can be written as

$$X_k = \frac{u_{jr} u_{kr}}{C_r \omega_{nr}} F_j \tag{6.7}$$

Where, ω_{mr} is the r-th natural frequency of the structure. It can be seen from equation (6.7) that to reduce the dynamic response of the structure, the corresponding natural frequency and damping can be increased. Many factors affect the damping, such as the connection of the structure and the lubricating oil characteristics of the mutually moving contact surfaces. When the structure is finished, the damping can be increased by adjusting the contact pressure. Therefore, the following mainly changes the structural parameters to increase the natural frequency, thereby reducing the dynamic response of the structure.

6.2.2 Average element energy analysis

Assuming that the r-th order feature vector is represented by ϕ_r, then equation (6.7) can be written as

$$X_k = \frac{\phi_{jr} \phi_{kr}}{C_r \omega_{nr}} F_j \tag{6.8}$$

ϕ_r chooses ф such that it satisfies $\phi_{jr}\ \phi_{kr} = 1$, then formula (6.8) becomes

$$X_k = \frac{1}{C_r \omega_{nr}} F_j \tag{6.9}$$

In the formula, $C_r = \phi_r^{\mathrm{T}} C \phi_r$. Differentiated equation (6.9) to any physical parameter q

$$\frac{\partial X_k}{\partial q} = -\frac{F_j}{C_r^2 \omega_{nr}^2} \left(C_r \frac{\partial \omega_{nr}}{\partial q} + \omega_{nr} \frac{\partial C_r}{\partial q} \right) \tag{6.10}$$

In the formula (see [167])

$$\frac{\partial \omega_{nr}}{\partial q} = -\frac{1}{2\omega_{nr} M_r} \left(\omega_{nr}^2 \phi_r^{\mathrm{T}} \frac{\partial M}{\partial q} \phi_r - \phi_r^{\mathrm{T}} \frac{\partial K}{\partial q} \phi_r \right) \tag{6.11}$$

$$\frac{\partial C_r}{\partial q} = 2\phi_r^{\mathrm{T}} \frac{\partial \phi_r}{\partial q} \tag{6.12}$$

$$\frac{\partial \boldsymbol{\phi}_r}{\partial q} = \sum_{m=1}^{N} \alpha_m \boldsymbol{\phi}_m \tag{6.13}$$

$$\alpha_m = -\frac{1}{2\boldsymbol{M}_r} \boldsymbol{\phi}_r^{\mathrm{T}} \frac{\partial \boldsymbol{M}}{\partial q} \boldsymbol{\phi}_r \quad (m = r) \tag{6.14}$$

$$\frac{\partial \boldsymbol{C}_r}{\partial q} = -\frac{\boldsymbol{C}_r}{\boldsymbol{M}_r} \boldsymbol{\phi}_r^{\mathrm{T}} \frac{\partial \boldsymbol{M}}{\partial q} \boldsymbol{\phi}_r \tag{6.15}$$

Substituting equation (6.11) and equation (6.15) into equation (6.10) gives

$$\frac{\partial \boldsymbol{X}_k}{\partial q} = -\frac{F_j}{\boldsymbol{C}_r \boldsymbol{M}_r \omega_{nr}^3} \left(\frac{1}{2} \boldsymbol{\phi}_r^{\mathrm{T}} \frac{\partial \boldsymbol{K}}{\partial q} \boldsymbol{\phi}_r - \frac{3}{2} \omega_{nr}^2 \boldsymbol{\phi}_r^{\mathrm{T}} \frac{\partial \boldsymbol{M}}{\partial q} \boldsymbol{\phi}_r \right) \tag{6.16}$$

The feature vector u_r is selected to be orthogonal to the mass matrix $\boldsymbol{M}$. Suppose

$$\boldsymbol{\phi}_r = \rho \boldsymbol{u}_r \tag{6.17}$$

Where ρ is a constant. Substituting equation (6.17) into equation (6.16), let $\boldsymbol{\beta}_r = \frac{\rho^2 F_j}{\boldsymbol{M}_r \boldsymbol{C}_r \omega_{nr}^3}$, we can get

$$\frac{\partial \boldsymbol{X}_k}{\partial q} = -\boldsymbol{\beta}_r \left(\frac{1}{2} \boldsymbol{u}_r^{\mathrm{T}} \frac{\partial \boldsymbol{K}}{\partial q} \boldsymbol{u}_r - \frac{3}{2} \omega_{nr}^2 \boldsymbol{u}_r^{\mathrm{T}} \frac{\partial \boldsymbol{M}}{\partial q} \boldsymbol{u}_r \right) \tag{6.18}$$

The physical parameter q may be spring stiffness, mass, beam cross-sectional area, and plate thickness. In this chapter, q represents the volume V_s of the substructure, then equation (6.18) becomes

$$\frac{\partial \boldsymbol{X}_k}{\partial V_s} = -\boldsymbol{\beta}_r \left(\frac{1}{2} \boldsymbol{u}_r^{\mathrm{T}} \frac{\partial \boldsymbol{K}}{\partial V_s} \boldsymbol{u}_r - \frac{3}{2} \omega_{nr}^2 \boldsymbol{u}_r^{\mathrm{T}} \frac{\partial \boldsymbol{M}}{\partial V_s} \boldsymbol{u}_r \right) \tag{6.19}$$

Suppose

$$\boldsymbol{E}_r = \frac{1}{2} \boldsymbol{u}_r^{\mathrm{T}} \boldsymbol{K} \boldsymbol{u}_r, \ \boldsymbol{T}_r = \frac{1}{2} \omega_{nr}^2 \boldsymbol{u}_r^{\mathrm{T}} \boldsymbol{M} \boldsymbol{u}_r \tag{6.20}$$

Substituting equation (6.20) into equation (6.19) gives

$$\frac{\partial \boldsymbol{X}_k}{\partial V_s} = -\boldsymbol{\beta}_r \frac{\partial \left(\boldsymbol{E}_r - 3\boldsymbol{T}_r \right)}{\partial V_s} \tag{6.21}$$

In the formula, E_r is the volume strain energy of the substructure; T_r is the kinetic energy of the substructure; V_s is the volume of the substructure; β_r is a positive constant.

In equation (6.21), $\frac{\partial(E_r - 3T_r)}{\partial V_s}$ represents the average element energy, including the average element volume strain energy and the average element kinetic energy. It can be seen from equation (6.21) that the partial derivative of the dynamic response with respect to the volume of the substructure is directly proportional to the average unit energy. The larger the average unit energy, the more sensitive the dynamic response.

6.3 Substructure's division

Structures include continuous structures (such as connecting rods, pistons, etc.) and discrete structures (such as truss structures, etc.). For structural optimization problems, the optimization model is very important, and the optimization model includes design variables and their value range, objective function, and constraint function. At present, engineering and technical personnel determine the objective function and the constraint function from the engineering needs, and comprehensively determine the design variables according to the geometric characteristics of the continuous structure, the relationship between the geometric elements, and the possibility of parameterization. When the number of design variables is large, you can use the sensitivity analysis method to delete design variables that are not sensitive to the objective function, thereby reducing the number of design variables, but doing so is time-consuming. For large-scale structures, it is very complicated to perform sensitivity analysis, and the value range of the reserved design variables is completely determined by the experience of engineering and technical personnel. The largest value range is often selected to ensure that the critical values are not missing.

The average element energy analysis of the substructure can initially determine the number of design variables and the range of design variables. The first problem to be solved is division of the substructure. If it is a continuous structure, it can be divided according to working conditions, structural shape, and parameterization, and then its design variables can be further determined according to the specific structural characteristics of each sub-structure.

The planar structure shown in Figure 6-1 can be divided into three sub-structures according to its geometric shape and parameterization, that is, the planar structure includes four parameters at the top, one rectangular hole at the middle, and one rectangular hole at the bottom. Namely, Sub_1, Sub_2, Sub_3, with each sub-structure having its own characteristics. Sub_1 has five design variables (A_1, B_1, C_1, D_1, E_1), Sub2 has four design variables (A_2, B_2, C_2, D_2), and Sub_3 also has four design variables (A_3, B_3, C3, D_3). After the sub-structure average element energy analysis, you can determine which design variables can be removed, which design variables should use the current value as the lower limit, and which design variables should use the current value as the upper limit, and then build an optimization model based in improving the efficiency of the process and simultaneously obtaining better values.

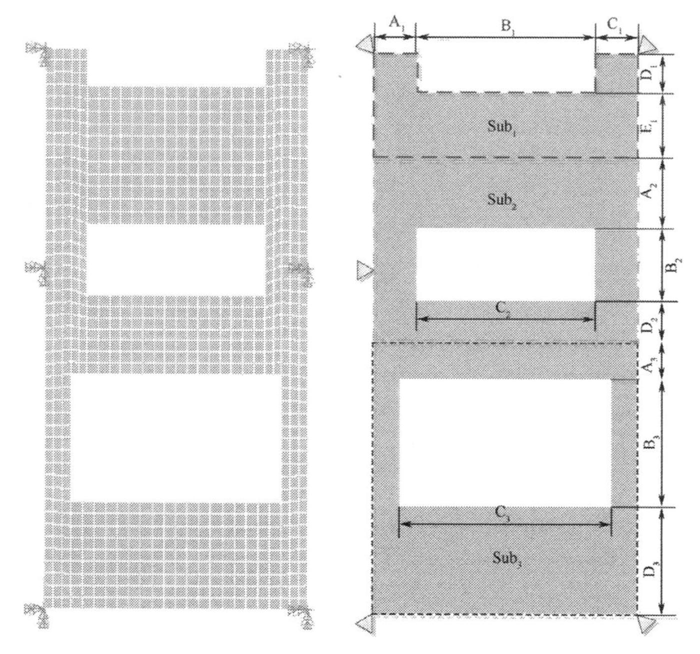

(a) The result of network classification (b) Division of the substructure

Figure 6-1 Plane structure

If it is a truss structure, each rod element can be regarded as a substructure, the average element volume strain energy and average element kinetic energy of each substructure can be calculated, and then the rod elements with similar values can be used as a substructure group. Then, the cross-sectional areas can be used as a design variable in each substructure group, so that solving the optimization model can greatly reduce the optimization scale and improve the optimization efficiency. As shown in Figure 6-2, the 18-bar plane truss structure can be divided into five groups if the sub-structure group is divided according to experience, that is, the upper horizontal elements 1, 4, 8, 12, 16 are one group, and the lower horizontal elements 6, 10, 4, 18 are another group, while vertical units 3, 7, 11, 15 are a group, left units 5, 9, 13, 17 are a group, right unit 2 is also a group. Taking the cross-sectional area of each set of rod elements as a design variable, the 18-bar planar truss structure optimization problem contains five design variables. In this way, the 18 design variables in the traditional optimization method are transformed into five design variables. From the perspective of optimization, the calculation time is greatly reduced. However, there is no theoretical basis for this transition process, and its rationality is unknown, and the value range of these five design variables is not clear and needs to be determined based on experience. To this end, this chapter proposes a solution from the perspective of the average element energy of the substructure, as shown in Table 6.1.

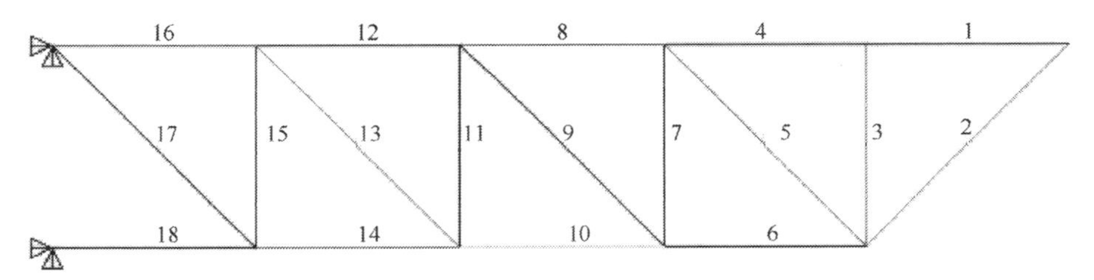

Figure 6-2 18-pole plane truss structure

Table 6.1 18 pole plane truss structure AUVSE-3AUVKE (Unit: 10^5 N/m^2)

Element number	AUVSE-3AUVKE	Element number	AUVSE-3AUVKE	Element number	AUVSE-3AUVKE
16	2.610885238	15	2.216582081	7	−2.015095248
18	2.35883264	9	−0.021305455	4	−2.553965162
17	1.227014281	11	−0.490030806	3	−4.22490324
13	0.881719452	8	−0.94713511	1	−4.282229688
12	0.66016753	5	−1.443409146	6	−5.316625814
14	0.580733555	10	−1.910251122	2	−9.178147295

Note: AUVSE-3AUVKE means average unit volume strain energy-average unit kinetic energy × 3;

AUVSE represents the average unit volume strain energy;

AUVKE represents the average unit kinetic energy.

It can be seen from Figure 6-3 and Table 6.1 that the average element kinetic energy of elements 1–11 is greater than the average element volume strain energy. Therefore, to increase the cross-sectional area and increase the natural frequency, the cross-sectional area of elements 1–11 can be used as a design variable, and we can use the current value as the lower limit. The average cell volume strain energy of units 12–18 is greater than the average unit kinetic energy. To reduce the cross-sectional area and increase the natural frequency, the cross-sectional area of units 12–18 can be used as another design variable and the current value can be used as its upper limit. In this way, the design variable of the optimization problem of the 18-bar plane truss structure is reduced to two by the average element energy analysis, and its value range is very clear.

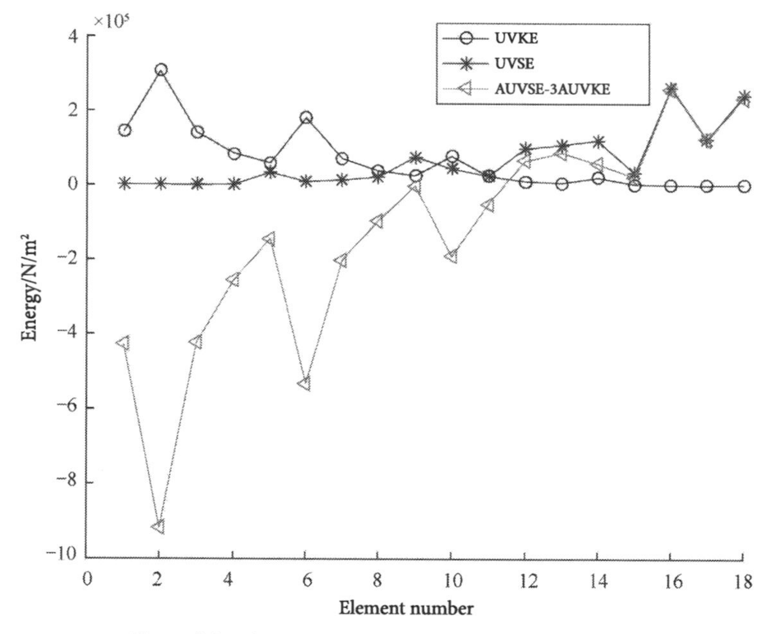

Figure 6-3 Average element energy of 18-bar plane truss

Note: UVKE means unit kinetic energy; UVSE means unit volume strain energy.

6.4 Implementation of structural dynamic characteristic optimization method based on substructure's average element energy

The optimization of the dynamic characteristics of a truss structure or a continuous structure including multiple substructures contains many design variables, which makes its sensitivity analysis very time-consuming. Therefore, this chapter proposes a structural dynamic characteristics optimization method based on the average element energy of substructures. As shown in 6.4, the specific steps are as follows.

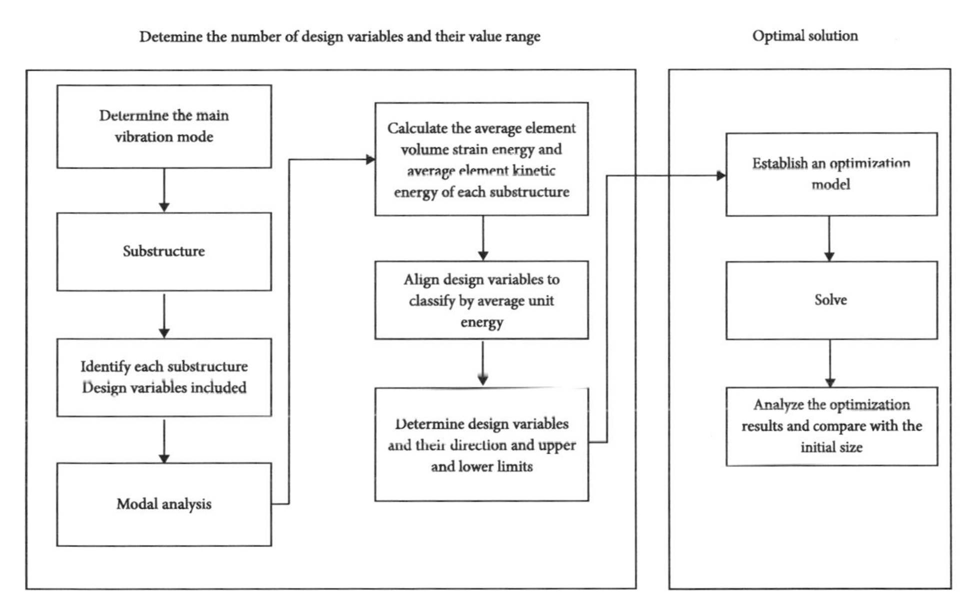

Figure 6-4 The optimization process

1) Determine the main vibration mode of the structure according to the vibration response. In general, the first few modes are taken as the main vibration modes. In actual engineering, the main vibration mode can be determined according to the actual stress of the structure and test analysis.
2) Divide the overall structure into sub-structures. Please refer to section 6.3 for the sub-structure division method of continuous structures and truss structures.
3) Determine the direction of the substructure's size change. When the whole structure vibrates in a particular main vibration mode, calculate the average element volume strain energy and average element kinetic energy of each substructure, and decide how to modify the design variables. For a substructure with a larger average element

volume strain energy and a smaller average element kinetic energy, the design variables it contains change in the direction of increasing stiffness. Meanwhile, for a substructure with a smaller average element volume strain energy and a larger average element kinetic energy, the values of the design variables change by decreasing in stiffness.

4) Establish optimization model. First, the three optimization models, such as the maximum natural frequency of the main vibration mode and the mass not exceeding the original nodal mass, are established, and the gradient-based optimizer must be used to solve the problem. Second, in the optimization model, the value range of the design variables must be dynamic, i.e. when the average unit volume strain energy of the substructure is relatively large, the lower limit of the size of the substructure must be the size of the atomic structure, and the upper limit can then be determined according to the feasibility of the geometric structure. On the other hand, when the average unit kinetic energy of the substructure is large, the upper limit of the size of the substructure must be the size of the atomic structure, and the lower limit can then also be determined according to the feasibility of the geometric structure. However, when the substructure reaches a very small size, it may be considered to delete the substructure completely.

6.5 Optimization design of the 124-bar truss structure

The 124-bar truss structure in Figure 6-5 has 49 hinges and 94 degrees of freedom. Its Elastic modulus is $E = 207$ GPa, Poisson's ratio is $\nu = 0.3$, density is $\rho = 7850$ kg/m^3, and cross-sectional area of the rod is 0.645×10^{-4} m^2. If each rod is used as a unit, there are 124 units in total (see Figure 6-5). The optimization goal is then to increase the first natural frequency as much as possible without increasing the volume (mass).

6.5.1 Using the methods mentioned in this chapter for optimization

Table 6.2 is the 124-bar truss structure AUVSE-3AUVKE. Assuming the first-order mode is the main vibration mode of the structure, using each bar as a sub-structure, the average unit energy of each sub-structure is obtained by formula (6.21) (see Figure 6-6), and then they are sorted from large to small, and then the substructures are divided with the geometric size of the design variables with a larger the average unit energy should be increased. This means that the current value is the lower limit of the design variable, while the geometric size of the design variable with a smaller average unit energy should be reduced, that is, the current value is the upper limit of the design variable, as shown in Table 6.3.

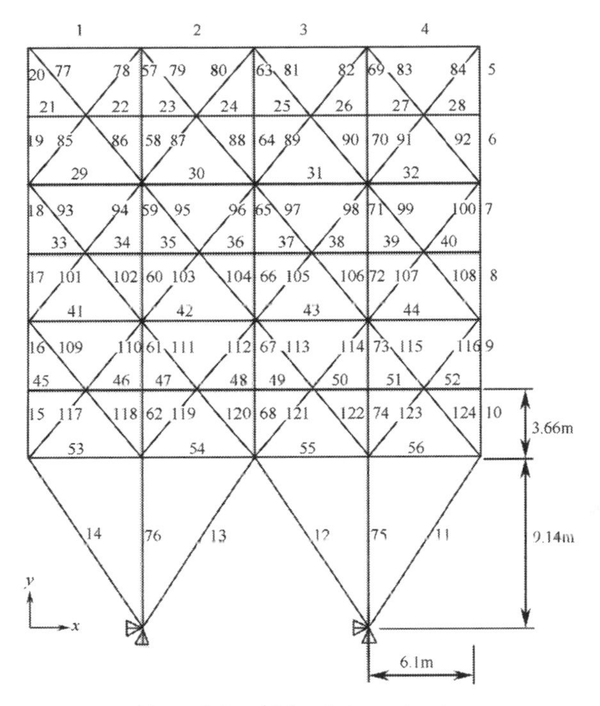

Figure 6-5 124 pole truss structure

Table 6.2 124 pole truss structure AUVSE-3AUVKE

Element number	AUVSE-3AUVKE	Element number	AUVSE-3AUVKE	Element number	AUVSE-3AUVKE
13	1.995681	118	−0.14293	43	−0.30156
12	1.995681	123	−0.14293	42	−0.30156
75	1.004119	47	−0.14657	9	−0.3024
76	1.004119	50	−0.14657	16	−0.3024
119	0.151567	111	−0.15893	104	−0.30601
122	0.151567	114	−0.15893	105	−0.30601
120	0.117884	51	−0.17714	109	−0.31775
121	0.117884	46	−0.17715	116	−0.31775
55	0.091938	49	−0.19118	41	−0.33812
54	0.091938	48	−0.19119	44	−0.33812
62	0.01785	10	−0.20472	66	−0.35825

(Continued)

Element number	AUVSE-3AUVKE	Element number	AUVSE-3AUVKE	Element number	AUVSE-3AUVKE
74	0.01785	15	−0.20472	60	−0.36338
11	0.017374	117	−0.22319	72	−0.36338
14	0.017374	124	−0.22319	102	−0.39546
61	−0.07991	67	−0.24108	107	−0.39546
73	−0.07991	110	−0.24655	101	−0.42178
112	−0.11867	115	−0.24655	108	−0.42178
113	−0.11867	45	−0.26607	37	−0.42553
53	−0.12188	52	−0.26607	36	−0.42555
56	−0.12188	103	−0.29328	35	−0.43275
68	−0.14158	106	−0.29328	38	−0.43275
8	−0.44317	29	−0.62666	21	−0.8077
17	−0.44317	32	−0.62666	28	−0.8077
96	−0.45839	88	−0.6369	80	−0.81498
97	−0.45839	89	−0.6369	81	−0.81498
39	−0.45992	87	−0.6475	63	−0.81658
34	−0.45995	90	−0.6475	79	−0.82792
95	−0.48588	64	−0.6488	82	−0.82792
98	−0.48588	58	−0.67198	57	−0.84017
65	−0.4948	70	−0.67198	69	−0.84017
33	−0.49966	86	−0.68611	78	−0.85555
40	−0.49966	91	−0.68611	83	−0.85555
59	−0.5015	85	−0.72281	77	−0.89311
71	−0.5015	92	−0.72281	84	−0.89311
94	−0.51883	25	−0.73337	3	−0.91282
99	−0.51883	24	−0.73341	2	−0.91282
93	−0.56995	23	−0.74452	5	−0.91574
100	−0.56995	26	−0.74452	20	−0.91574
31	−0.57727	6	−0.74851	1	−0.96264
30	−0.57727	19	−0.74851	4	−0.96264
7	−0.58002	27	−0.7705		
18	−0.58002	22	−0.77054		

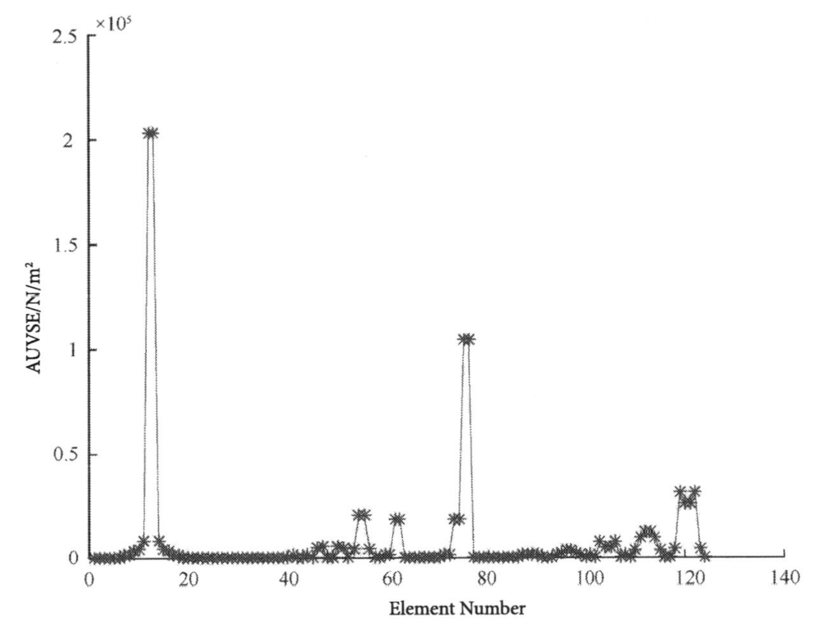

(a) Average unit volume strain energy

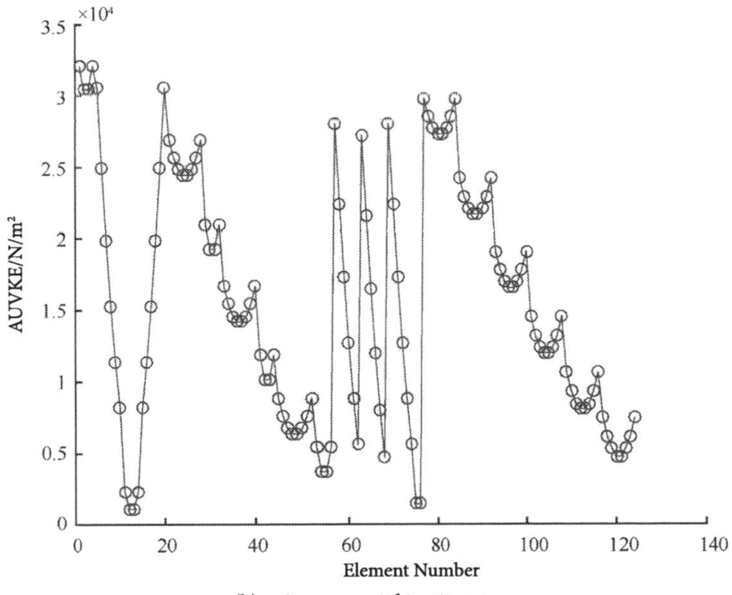

(b) Average unit kinetic energy

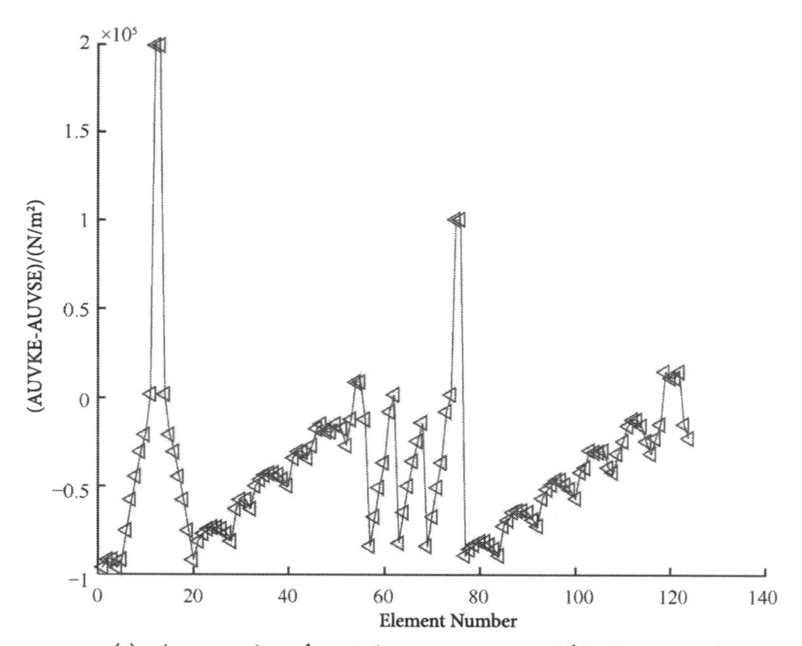

(c) Average unit product strain energy-average unit kinetic energy × 3

Figure 6-6 Average element energy of 124-bar truss structure

Table 6.3 Selection of design variables

Design variable	Element number	Lower limit	Upper limit
X_{DV1} (4)	12, 13, 75, 76	6.45E-05	6.45E-04
X_{DV2} (6)	54, 55, 119–122	6.45E-05	6.45E-04
X_{DV3} (4)	11, 14, 62, 74	6.45E-05	6.45E-04
X_{DV4} (57)	8–10, 15–17, 34–39, 41–53, 56, 60, 61, 66–68, 72, 73, 95–98, 101–118, 123, 124	0	6.45E-05
X_{DV5} (53)	1–7, 18–33, 40, 57–59, 63–65, 69–71, 77–94, 99, 100	0	6.45E-05

The optimization model is optimized according to formula (6.1) and formula (6.2), and the obtained results are shown in Figure 6-7, Table 6.4 and Table 6.5. It can be seen that whether formula (6.1) or formula (6.2) is used as the optimization model, both can be easily converted. For the optimization with the smallest volume as the goal, the volume is reduced by 92.74%. For the optimization with the largest first natural frequency as the goal, the first natural frequency is increased by 143.38%, and its effect is obvious. The robustness of the optimization model is very good, the optimization process is not affected by the initial conditions, indicating that the method of selecting design variables and determining the range of values proposed in this chapter is correct and effective.

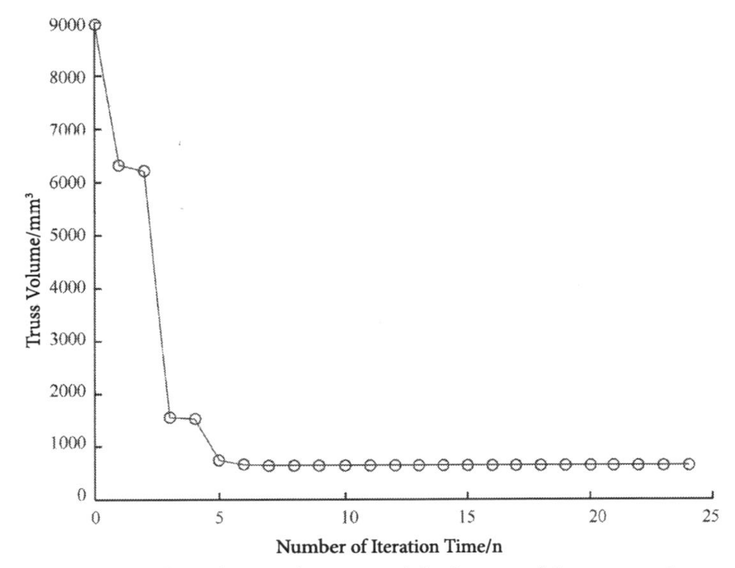

(a) Taking the smallest volume as the target and the first natural frequency as the constraint

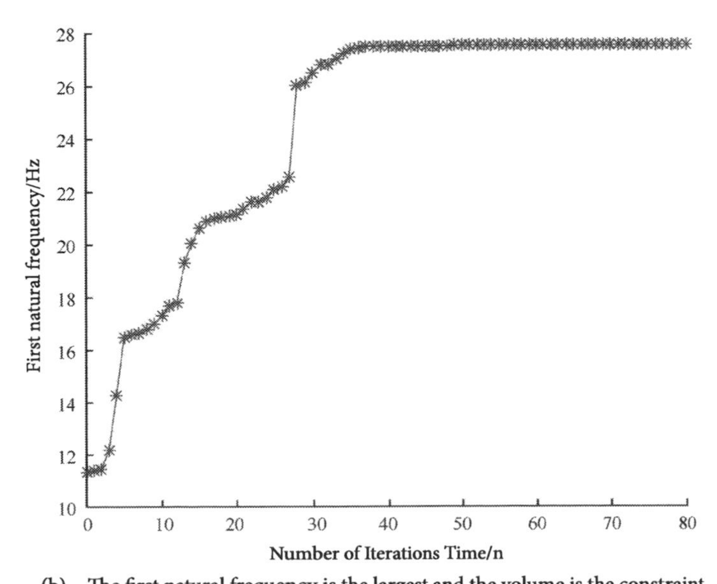

(b) The first natural frequency is the largest and the volume is the constraint

Figure 6-7 Optimization results obtained by the method mentioned in this chapter

Table 6.4 Design variables optimized by using two models (1) (Unit: $10^{-5}m^2$)

Design variable	The optimization model corresponding to formula (6.1)		Equation (6.2) corresponding optimization model	
	Initial value	The optimal value	Initial value	The optimal value
X_{DV1}	6.45000000	6.45000497	6.45000000	64.37565467
X_{DV2}	6.45000000	6.4500064	6.45000000	20.20888033
X_{DV3}	6.45000000	6.45000678	6.45000000	6.4630552
X_{DV4}	6.45000000	0.00511895	6.45000000	0.44175131
X_{DV5}	6.45000000	0.00000225	6.45000000	0.00655862

Table 6.5 Target values and time-consuming optimized by using two models (1)

The optimization model corresponding to formula (6.1)		Equation (6.2) corresponding optimization model	
Minimum volume/ mm^3	651.4698244	Maximum frequency/ Hz	27.54793635
Time-consuming/s	331.417	Time-consuming/s	1148.788
Initial volume/mm^3	8972.596	Initial frequency/Hz	11.31876

6.5.2 Determining the design variables and their value ranges based on geometric characteristics and experience

If the traditional method is used to divide the 124 elements of the truss structure into 6 groups as shown in Table 6.6 according to geometric characteristics, and each group of elements is used as a design variable, the optimization model can be carried out using equations (6.1) and (6.2). The solution and the results are shown in Figure 6-8, Table 6.7 and Table 6.8. It can be seen that the optimization with the smallest volume as the goal reduces the volume by 88.12%. For the optimization with the largest first natural frequency as the goal, the first natural frequency increases by 150.38%.

Table 6.6 124 Variable truss design with variable grouping

Design variable	Corresponding unit
X_{DV1} (tilted to the left, 24 units)	77, 79, 81, 83, 86, 88, 90, 92, 93, 95, 97, 99, 102, 104, 106, 108, 109, 111, 113, 115, 118, 120, 122, 124
X_{DV2} (tilted to the right, 24 units)	78, 80, 82, 84, 85, 87, 89, 91, 94, 96, 98, 100, 101, 103, 105, 107, 110, 112, 114, 116, 117, 119, 121, 123
X_{DV3} (horizontal long pole, 16 units)	1–4, 29–32, 41–44, 53–56
X_{DV4} (horizontal short bar, 24 units)	21–28, 33–40, 45–52
X_{DV5} (vertical short bar, 30 units)	5–10, 15–20, 57–74
X_{DV6} (support rod, 6 units)	11–14, 75, 76

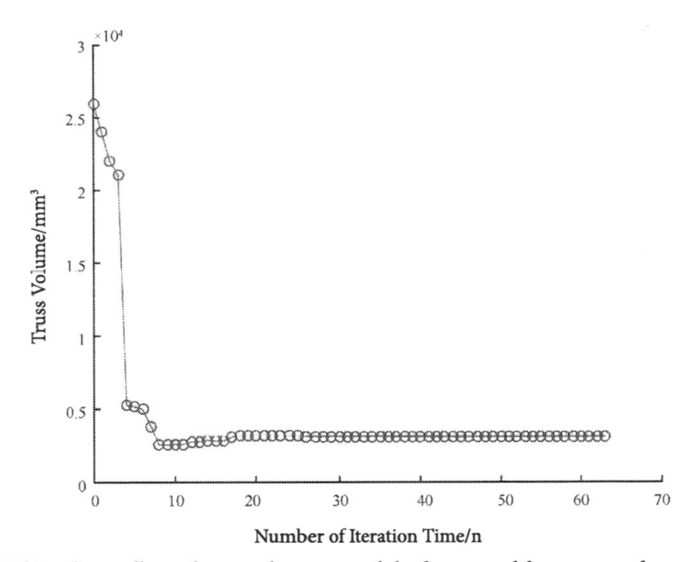

(a) Taking the smallest volume as the target and the first natural frequency as the constraint

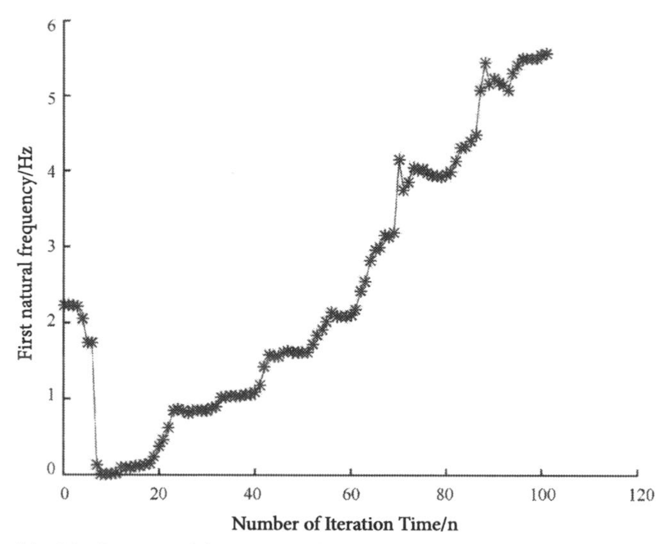

(b) The first natural frequency is the largest and the volume is the constraint

Figure 6-8 Optimization results obtained using traditional methods

Table 6.7 Design variables optimized by using two models (2) (Unit: $10^{-5}m^2$)

Design variable	The optimization model corresponding to formula (6.1)		Equation (6.2) corresponding optimization model	
	Initial value	The optimal value	Initial value	The optimal value
X_{DV1}	6.45000000	6.45000135	6.45000000	6.49858160
X_{DV2}	6.45000000	6.45000139	6.45000000	6.49911855
X_{DV3}	6.45000000	6.45000157	6.45000000	6.50380443
X_{DV4}	6.45000000	0.03744709	6.45000000	0.13842305
X_{DV5}	6.45000000	5.23031807	6.45000000	6.43493111

Table 6.8 Target values and time consumption optimized by using two models (2)

The optimization model corresponding to formula (6.1)		Equation (6.2) corresponding optimization model	
Minimum volume/mm^3	3082.97542727	Maximum frequency/Hz	5.5679342
Time-consuming/s	902.212	Time-consuming/s	1200.012
Initial volume/mm^3	25954.04	Initial frequency/Hz	2.223775

For the results obtained by using the method proposed in this chapter for optimization, only the percentage of change does not reveal the advantages of the method presented in this chapter. However, if the three aspects of percentage change, optimization iteration process, and optimization results are combined, the optimization result of the method proposed in this chapter is obviously superior. In terms of optimization results, the minimum volume obtained by the method proposed in this chapter is 651.4469824437 mm^3, and the traditional method obtained is 3082.97542727 mm^3, meaning the latter is 4.7323 times more accurate than the former. The maximum first natural frequency obtained by the method proposed in this chapter is 27.54793635 Hz, while the traditional method obtains 5.5679342 Hz, and the former is 4.9476 times higher than that of the latter, and the latter has very poor convergence in the optimized iterative process with the first natural frequency as the maximum goal. Taken together, the method proposed in this chapter is very effective for optimization.

6.5.3 Using the optimization model of (6.3) to optimize and compare two methods

In engineering, sometimes the natural frequency of the main vibration mode is required to reach the maximum speed when the volume is constant. Equation (6.3) is the expression of this requirement. Using the method proposed in this chapter and the traditional method to solving the Equation (6.3) and by comparing the two, the results are shown in Figure 6-9, Table 6.9 and Table 6.10. To enhance comparability, the two methods use the same optimized configuration, including initial values, convergence accuracy, and so on. It can be seen from Figure 6-9 (a) that the first natural frequency obtained by the method proposed in this chapter at the 24th iteration is 27.08437 Hz, which is very close to the optimal solution of 27.45954 Hz. Meanwhile, the traditional method does not converge the same way due to the design variables because the change

Table 6.9 Design variables optimized by two methods (Unit: $10^{-5}m^2$)

Design variable	The optimization model corresponding to formula (6.1)		Equation (6.2) corresponding optimization model	
	Initial value	The optimal value	Initial value	The optimal value
X_{DV1}	6.45000000	64.49734549	6.45000000	7.08958470
X_{DV2}	6.45000000	24.99877391	6.45000000	7.79082535
X_{DV3}	6.45000000	6.44079153	6.45000000	7.03782795
X_{DV4}	6.45000000	0.48945591	6.45000000	2.66241225
X_{DV5}	6.45000000	0.00722838	6.45000000	6.44668309
X_{DV6}			6.45000000	6.35036798

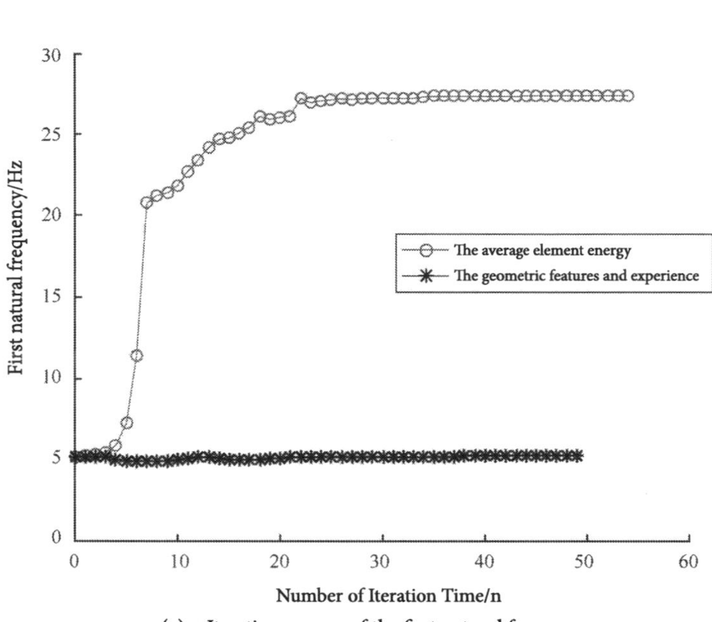

(a) Iterative process of the first natural frequency

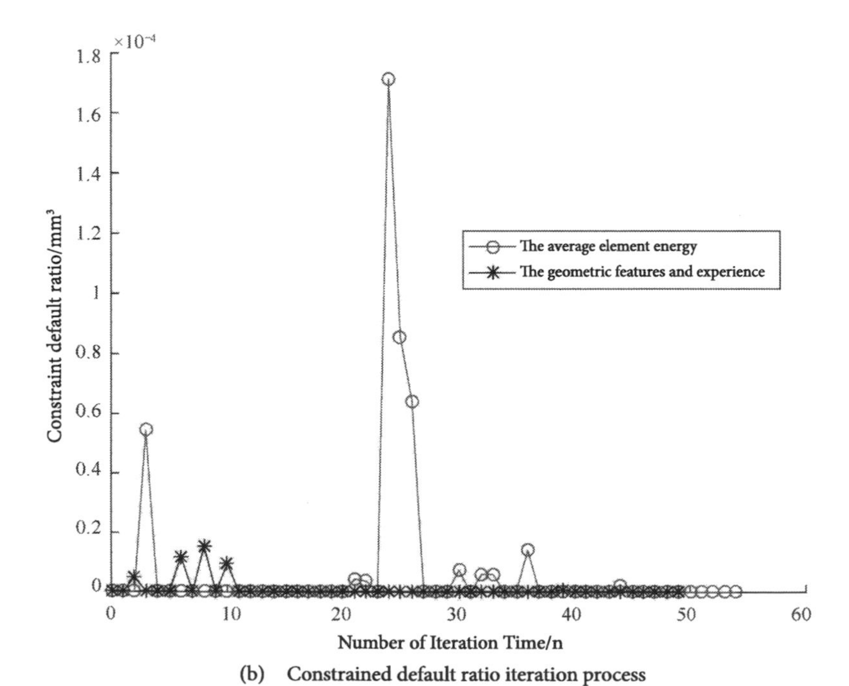

(b) Constrained default ratio iteration process

Figure 6-9 Optimization iteration process with volume as the constraint and the first natural frequency as the maximum goal

of the design variables is already less than the convergence accuracy, and its iteration stops. Therefore, in this case, it is meaningless to compare the number of iterations and its time-consumption. The method proposed in this chapter improves the first natural frequency by 423.27% while keeping the volume unchanged, and obtains the constraint default ratio at the 55th iteration as 7.841×10^{-15} mm^3, which shows that the method proposed in this chapter is correct and effective.

From the theoretical analysis of this chapter, because the average unit volume strain energy of the substructures corresponding to X_{DV1} and X_{DV2} is relatively large, in order to reduce the dynamic response, the geometric dimensions of the substructures corresponding to X_{DV1} and X_{DV2} should be increased. It can be seen from Table 6.9 that these two design variables have indeed increased, and are consistent with the theoretical analysis. The average unit kinetic energy of the substructures corresponding to X_{DV4} and X_{DV5} is relatively large. In order to reduce the dynamic response, the geometric dimensions of the corresponding substructures should be reduced. It can be seen from Table 6.9 that the theoretical analysis is consistent with the actual calculation. The average cell volume strain energy of the X_{DV3} cell is very close to the average cell kinetic energy, so it can be used for constrained optimization with constant volume. This can be seen from Table 6.9 and it is also reasonable conclusion.

Table 6.10 Using two methods to optimize the target value and time-consuming

The methods mentioned in this chapter		Traditional method	
Maximum frequency/Hz	27.45954	Maximum frequency/Hz	5.2725135
Sending times/time	55	Sending times/time	50
Time-consuming/s	647.8536	Time-consuming/s	721.915
Initial frequency/Hz	5.24767	Initial frequency/Hz	5.24767

6.6 Summary of this chapter

In this chapter, based on the relationship between the average element volume strain energy and the average element kinetic energy, as well as the dynamic response of the structure, the substructure composed of a certain number of elements and geometric features was used as a design variable. Then, the geometric size was increased according to the substructure with a larger average element volume strain energy. The sub-structure with a larger average unit kinetic energy was reduced in size according to the geometric theory, which greatly reduced the value range of the design variables, and ultimately improved the performance of the structural dynamic characteristics optimization, and obtained the following conclusions:

1) The size of the average element volume strain energy and the average element kinetic energy can be used to determine the contribution of the element or sub-structure (multiple element sets) to the overall dynamic response of the structure, thereby determining the design variables and their value ranges. For example, for the 124-bar truss structure, the elements with an average element volume strain energy of an order of magnitude were used as a group, and set as a design variable. The value range was limited to the initial value. The size of the substructure corresponding to X_{DV1} and X_{DV2} needed to be increased. The sub-structure corresponding to X_{DV3} had to be maintained or even deleted, while the size of the sub-structure corresponding to X_{DV4} and X_{DV5} needed to be reduced.
2) Taking the smallest volume as the optimization goal and the first natural frequency as the constraint condition, the volume that could be obtained after optimization was very small, and the first natural frequency was not decreased. However, the load that this structure could withstand in practice is very small, which makes little sense. For example, when the minimum volume is 651.4469824437 mm^3, the minimum cross-sectional area of the 124-bar truss structure is 2.25×10^{-11} mm^2; and the first natural frequency is the maximum optimization goal, and the volume is therefore not the constraint condition, while more reasonable optimization results could still be obtained. For example, when the minimum cross-sectional area of the 124-bar truss structure is 7.22838×10^{-8} mm^2, the natural frequency of the main vibration mode is increased by 423.27%.
3) The structural dynamic characteristic optimization model is the key factor for optimization convergence and obtaining reasonable optimization results. The optimization model based on the sub-structure average unit energy method for optimization can converge well and the optimization results are therefore reasonable, but the traditional method is difficult to converge.

CHAPTER 7

A Structural Dynamic Reduction Method Based on Node Ritz Potential Energy and Principal Degree of Freedom

A large number of degrees of freedom models need to be calculated for dynamic analysis of complex structures, but the applied high-frequency excitation force requires very small calculation steps. This will lead to an exponential increase in calculation time. In order to improve the calculation efficiency, a large number of degrees of freedom models can be replaced with a small number of degrees of freedom models, which means that when the model is reduced, it can still ensure a certain accuracy. The so-called model reduction means that through a particular transformation, the sub-degrees of freedom which have little influence on the overall structural dynamics analysis can be expressed by a small number of degrees of freedom which have a large influence on the overall structural dynamics analysis. This will achieve the main aim of reducing the degrees of freedom. Among them, a few degrees of freedom will be the master degrees of freedom. However, choosing the main degree of freedom from the huge array of degrees of freedom is still a very challenging problem in the field of structural dynamics. At the same time, the academic community currently proposes some principles for choosing the main degree of freedom, the most representative of which are first, the structural vibration direction that is set as the main degree of freedom or second, the mass or rotational inertia that is relatively large and which stiffness is relatively small can be selected as the main degree of freedom, or third, selecting the main degree of freedom at the position where force or non-zero displacement is applied. These principles are only guiding ideology. When choosing the main degree of freedom, there is more randomness to the process. The position and number of master degrees of freedom directly affect the modal analysis and reduce the accuracy of the mass matrix. The sequence element elimination method [168] is the most effective method for selecting the principal degree of freedom, which takes the larger of the diagonal elements of the structural mass matrix and the stiffness matrix as the principal degree of freedom. Matta et al. [169] adjusted the remaining degrees of freedom to compensate for the effects of each eliminated degree of freedom. Luo Hong et al. [170] proposed two methods for selecting the master degrees of freedom, and analyzed the characteristics and applicable scope of these two methods with single-layer cantilever beam as the object. Liu Xiaobao et al. [171] conducted a variety of studies on the selection of master degrees of freedom, and showed that the largest influencing factor of errors in static reduction mode analysis of structures is the number and distribution of the master degrees of freedom, especially the master degrees of freedom in distribution. Bao Xuehai et al. [172] took the bogie as the analysis object and proposed some criteria for selecting the master degrees of freedom.

After selecting the master degrees of freedom, the next step is to reduce the model. The earliest model reduction method proposed by Guyan [173] is called Guyan reduction method, also known as static reduction method. This method ignores the inertia and damping terms related to the degrees of freedom, and it divides the mass matrix, the stiffness matrix, the state vector and the load vector into two master degrees of freedom and second degrees of freedom, and then undergoes matrix transformation to include the master degrees of freedom. After considering the inertial term, an improved model reduction method is produced. The inertial term is obtained through the Guyan reduction method, and the result is closer to the overall structural mode. [174, 175] In the modal superposition method, the required modes to be superimposed have absolutely no relationship with the load, but in essence, some modes may have a small contribution, so you can consider using the Ritz vector superposition, because the Ritz vector can reflect the dynamic characteristics of the structure. [176, 177] They include the indicates containing the second degree of freedom. Aiming at the shortcomings of Guyan reduction method, an improved idea and derived the relevant formula were proposed. [179] The substructure method for dynamic analysis was proposed by combining the superposition of Ritz vectors with the static substructure method. [142] In the damage identification of engineering structures, the results with the highest recognition accuracy were obtained by applying the stepwise approximation model polycondensation method and adopted the first-order polycondensation model of the improved Guyan recursive polycondensation method. [180] The DC gain of each mode of the structure is used as the criterion for value judgment, and it cuts off the modes and realizes the model reduction. [181, 182] Rixena et al. [183] proposed the use of low-order polynomials or piecewise Lagrange multiplier polynomials to connect independent dynamic reduction sub-model, and also proposed a smoothing method based on the Rayleigh-Ritz method.

It can be seen from the literature analysis that at present, a lot of research has been carried out on the structural dynamic reduction method and the main degree of freedom selection method, which impacts it the most. However, there is no accurate and high-precision main degree of freedom selection method. Based on this, this chapter proposes a method for dynamic reduction of structures based on the master degrees of freedom of the node's Leeds potential energy. First, the node's Leeds potential energy is defined by using the characteristics of the Leeds vector associated with the dynamic characteristics of the structure and the load distribution form the structure bears. Second, on this basis, the formula for calculating the weighting coefficient is given, and then the node Ritz potential energy is multiplied by the weighting coefficient to obtain the node Ritz potential energy vector, and the master degrees of freedom are selected based on it. The small-scale structural dynamic equations are therefore obtained by the improved dynamic reduction method and solved.

7.1 Calculation of node Leeds potential energy and selection of master degrees of freedom

The modal superposition method is an effective method for calculating the dynamic response of a structure. It uses the orthogonality of the vibration modes to decouple the dynamic equations, and then solve each equation separately to superimpose it. However, some of the modals involved in the calculations have a small effect on the dynamic responses, because the vibration mode has no relationship with the load distribution of the structure. The Ritz vector is a set of orthogonal vectors related to the spatial distribution of the load. It is obtained by normalizing and orthogonalizing the initial response and the mass matrix, and it can then reflect the influence of the inertial force. The construction process of the Ritz vector is described as follows:

First, the initial vector $\boldsymbol{x}_1$ is obtained by normalizing $\boldsymbol{M}$, as shown in equation (7.1).

$$\boldsymbol{K}\boldsymbol{x}_1^* = \boldsymbol{M}_{ii},\ \boldsymbol{x}_1^{\mathrm{T}}\boldsymbol{M}\boldsymbol{x}_1 = 1 \tag{7.1}$$

Then an iterative formula is constructed as shown in equation (7.2).

$$\boldsymbol{x}_{i+1} = \tilde{\boldsymbol{x}}_{i+1} \Big/ \sqrt{\tilde{\boldsymbol{x}}_{i+1}^{\mathrm{T}}\boldsymbol{M}\tilde{\boldsymbol{x}}_{i+1}} \tag{7.2}$$

In the formula,

$$\tilde{\boldsymbol{x}}_{i+1} = \boldsymbol{x}_{i+1}^* - \sum_{j=1}^{i} c_{ji} \cdot \boldsymbol{x}_j \tag{7.3}$$

$$c_{ji} = \boldsymbol{x}_j^{\mathrm{T}}\boldsymbol{M}\boldsymbol{x}_{i+1}^*\ (j = 1, 2, \cdots, i) \tag{7.4}$$

$$x^*_{i+1} = K^{-1}Mx_i \tag{7.5}$$

The Ritz vector corresponds to the modal vector. Take the degrees of freedom corresponding to the P_1, P_2, and P_k, and the largest Ritz vector components in the first k-order Ritz vectors, combine them, and delete the coincidence to obtain the final main degree of freedom – this is called the main degree of freedom selection method based on the Ritz vector.

The nodal Ritz potential energy is obtained by multiplying the Ritz vector and the mass vector point of the nodal degree of freedom (the precondition is to convert the modal space to the Ritz vector space). The elements of each row of the mass matrix are summed as the DOF quality of structural finite element nodes. The degree of freedom corresponding to the larger component of the node Ritz potential energy is selected as the main degree of freedom. However, this will overemphasize the low-order frequencies and ignore the high-order frequencies. Therefore, the problem of excessive concentration of the calculation results on the low-order frequencies can be solved by defining weighting coefficients.

The calculation formula of node Ritz potential energy is as follows:

$$P^i_{\text{Ritz}} = w^i_N \left\{ x_i \sum_{j=1}^{n} M_{ij} \right\}_i (i = 1, 2, \cdots, n) \tag{7.6}$$

In the formula, w^i_N is a weighting coefficient, expressed as follows:

$$w^i_N = \left(\frac{x_i}{x_{\max}} \right)^2 \tag{7.7}$$

7.2 Constructing a reduction system

The eigenvalue problem of linear finite element structure can be expressed as

$$K\varphi = \omega^2 M\varphi \tag{7.8}$$

In the formula, K is the nth order stiffness matrix; M is the nth order mass matrix; ω is the natural frequency; φ is the eigenvector. Equation (7.8) can be decomposed into

$$\begin{bmatrix} \boldsymbol{K}_{pp} & \boldsymbol{K}_{ps} \\ \boldsymbol{K}_{sp} & \boldsymbol{K}_{ss} \end{bmatrix} \begin{bmatrix} \boldsymbol{\varphi}_p \\ \boldsymbol{\varphi}_s \end{bmatrix} = \omega^2 \begin{bmatrix} \boldsymbol{M}_{pp} & \boldsymbol{M}_{ps} \\ \boldsymbol{M}_{ps} & \boldsymbol{M}_{ss} \end{bmatrix} \begin{bmatrix} \boldsymbol{\varphi}_p \\ \boldsymbol{\varphi}_s \end{bmatrix} \tag{7.9}$$

In the formula, φ_p is the eigenvector corresponding to the main degree of freedom; φ_s is the eigenvector corresponding to the secondary degree of freedom; the subscript p indicates the quantity corresponding to the main degree of freedom; the subscript s indicates the quantity corresponding to the secondary degree of freedom.

$$\begin{bmatrix} \boldsymbol{\varphi}_p \\ \boldsymbol{\varphi}_s \end{bmatrix} = \boldsymbol{T}\boldsymbol{\varphi}_p \tag{7.10}$$

$$\bar{\boldsymbol{M}} = \boldsymbol{T}^{T}\boldsymbol{M}\boldsymbol{T}, \bar{\boldsymbol{K}} = \boldsymbol{T}^{T}\boldsymbol{K}\boldsymbol{T} \tag{7.11}$$

In the formula, $\bar{\boldsymbol{M}}$ is the reduced mass matrix; $\bar{\boldsymbol{K}}$ is the reduced stiffness matrix; $\boldsymbol{T}$ is the conversion matrix.

The steady-state harmonic response of the structure can be expressed as

$$\left(\begin{bmatrix} \boldsymbol{K}_{pp} & \boldsymbol{K}_{ps} \\ \boldsymbol{K}_{sp} & \boldsymbol{K}_{ss} \end{bmatrix} - \Omega^2 \begin{bmatrix} \boldsymbol{M}_{pp} & \boldsymbol{M}_{ps} \\ \boldsymbol{M}_{ps} & \boldsymbol{M}_{ss} \end{bmatrix} \right) \begin{bmatrix} \boldsymbol{x}_p \\ \boldsymbol{x}_s \end{bmatrix} = \begin{bmatrix} \boldsymbol{f}_p \\ \boldsymbol{f}_s \end{bmatrix} \tag{7.12}$$

In the formula, Ω is the frequency of the harmonic excitation force; f_p, f_s are excitation force.

If $f_s = 0$, the exact reduction relationship can be obtained from equation (7.12):

$$\boldsymbol{x}_s = \left(\boldsymbol{I} - \Omega^2 \boldsymbol{K}_{ss}^{-1}\boldsymbol{M}_{ss}\right)^{-1} \left(-\boldsymbol{K}_{ss}^{-1}\boldsymbol{K}_{sp} + \Omega^2 \boldsymbol{K}_{ss}^{-1}\boldsymbol{M}_{sp}\right)\boldsymbol{x}_p \tag{7.13}$$

Apply the binomial theorem to expand the equation (7.13) and omit the terms above second order, we can get:

$$\boldsymbol{x}_s = \left[-\boldsymbol{K}_{ss}^{-1}\boldsymbol{K}_{sp} + \Omega^2 \left(\boldsymbol{K}_{ss}^{-1}\boldsymbol{K}_{sp} - \boldsymbol{K}_{ss}^{-1}\boldsymbol{M}_{ss}\boldsymbol{K}_{ss}^{-1}\boldsymbol{K}_{sp}\right)\right]\boldsymbol{x}_p \tag{7.14}$$

When $\Omega = 0$,

$$\boldsymbol{x}_s = \left[-\boldsymbol{K}_{ss}^{-1}\boldsymbol{K}_{sp}\right]\boldsymbol{x}_p \tag{7.15}$$

Then, the static reduction conversion matrix is

$$\boldsymbol{T}_{stat} = \begin{bmatrix} \boldsymbol{I} \\ -\boldsymbol{K}_{ss}^{-1}\boldsymbol{K}_{sp} \end{bmatrix} \tag{7.16}$$

Apply the following approximation to eliminate Ω:

$$\Omega^2 \boldsymbol{x}_{\mathrm{p}} = \boldsymbol{D}_{\mathrm{stat}} \boldsymbol{x}_{\mathrm{p}} \tag{7.17}$$

$$\boldsymbol{D}_{\mathrm{stat}} = \left(\boldsymbol{T}_{\mathrm{stat}}^{\mathrm{T}} \boldsymbol{M} \boldsymbol{T}_{\mathrm{stat}}\right)^{-1} \left(\boldsymbol{T}_{\mathrm{stat}}^{\mathrm{T}} \boldsymbol{K} \boldsymbol{T}_{\mathrm{stat}}\right) \tag{7.18}$$

$$\boldsymbol{x}_{\mathrm{s}} = \left[-\boldsymbol{K}_{\mathrm{ss}}^{-1}\boldsymbol{K}_{\mathrm{sp}} + \left(\boldsymbol{K}_{\mathrm{ss}}^{-1}\boldsymbol{K}_{\mathrm{sp}} - \boldsymbol{K}_{\mathrm{ss}}^{-1}\boldsymbol{M}_{\mathrm{ss}}\boldsymbol{K}_{\mathrm{ss}}^{-1}\boldsymbol{K}_{\mathrm{sp}}\right)\boldsymbol{D}_{\mathrm{stat}}\right]\boldsymbol{x}_{\mathrm{p}} \tag{7.19}$$

$$\boldsymbol{T}_{\mathrm{IRS}} = \begin{bmatrix} \boldsymbol{I} \\ -\boldsymbol{K}_{\mathrm{ss}}^{-1}\boldsymbol{K}_{\mathrm{sp}} + \left(\boldsymbol{K}_{\mathrm{ss}}^{-1}\boldsymbol{K}_{\mathrm{sp}} - \boldsymbol{K}_{\mathrm{ss}}^{-1}\boldsymbol{M}_{\mathrm{ss}}\boldsymbol{K}_{\mathrm{ss}}^{-1}\boldsymbol{K}_{\mathrm{sp}}\right)\boldsymbol{D}_{\mathrm{stat}} \end{bmatrix} \tag{7.20}$$

7.3 Implementation of structural dynamic reduction method based on the master degrees of freedom of node Ritz potential energy

Figure 7-1 shows the flow of the structure dynamic reduction method based on the master degrees of freedom of Leeds potential energy node. Its implementation steps are as follows:

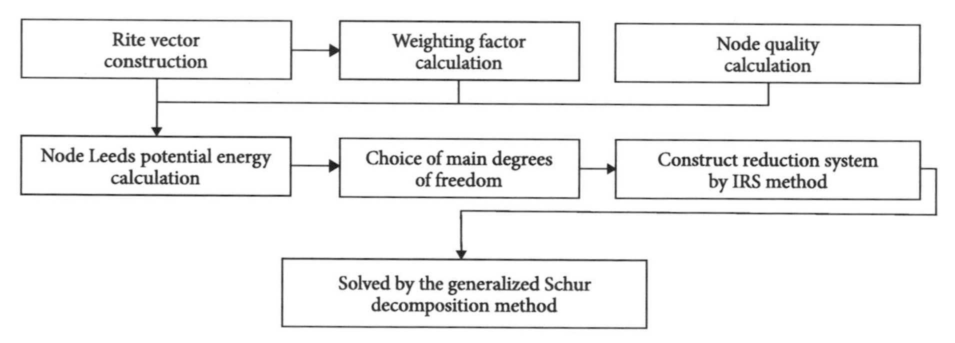

Figure 7-1 Process flow of structural dynamic reduction method based on the master degrees of freedom of node Ritz potential energy

Step 1: Construct the Ritz vector. The Ritz vector is constructed by formula (7.1)–formula (7.5). The first Ritz vector is obtained by multiplying the diagonal elements of the mass matrix and the inverse matrix of the stiffness matrix, and by normalizing the result with the mass matrix.

Step 2: Calculate the node Ritz potential energy. The node Ritz potential energy is the product of the node's mass and the corresponding component of the Leeds vector, which represents the contribution rate of the dynamic characteristics of the structure. The larger one contributes more, and the smaller one contributes less. However, if there is too much emphasis on low-order frequencies, the accuracy of high-order frequencies can be improved by defining weighting coefficients.

Step 3: Main freedom degree selection. Calculate the node Ritz potential energy through equation (7.6), and select the one with the larger component as the main degree of freedom.

Step 4: Construct a reduction system. Through the IRS method, based on the static reduction method, the structural inertia force is considered. The inertial force of the structure can make the modal results closer to the modal of the complete model. Equation (7.20) is the most important conversion matrix, and Equation (7.8) – Equation (7.19) are its derivation processes.

Step 5: Solve it using the generalized Schur decomposition method. [184] For any n-order matrix A, there is a unitary matrix U such that $U'AU$ becomes an upper triangular matrix, and the diagonal elements of the upper triangular matrix are the eigenvalues of A. This property can be used to solve the reduced system.

7.4 Example analysis

7.4.1 Cylindrical curved plate

The radius of the cylindrical curved plate is 100 mm, the height is 100 mm, and both sides are fixed, as shown in Figure 7-2. The elastic modulus is 0.3 MPa, the Poisson's ratio is 0.3, and the density is 0.01 kg/m³. It uses SHELL63 cells for meshing, and a total of 216 cells, 247 nodes, and 1326 degrees of freedom are obtained.

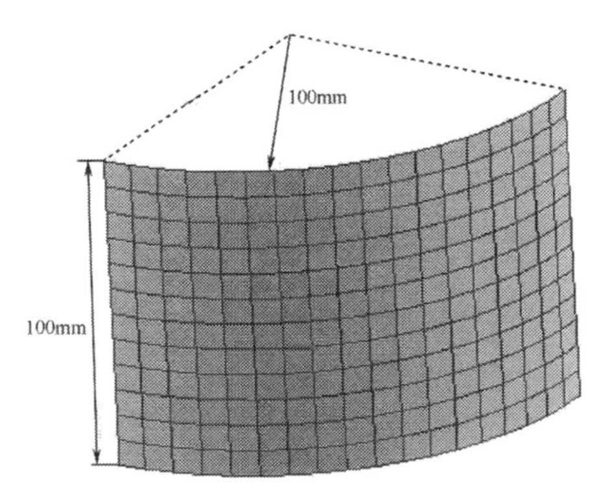

Figure 7-2 Geometric parameters of cylindrical curved plate

The master degrees of freedom are selected based on the Ritz vector method and the node Ritz potential energy method, and the elements connected to the master degrees of freedom are displayed as shown in Figure 7-3. Among them, the darker color is the unit corresponding to the master degrees of freedom. It can be seen from Figure 7-3 that the master degrees of freedom selected based on the Ritz vector method are too concentrated at the center of the structure, which is an important manifestation of excessive emphasis on low-order frequencies. Meanwhile, the master degrees of freedom selected based on the node Ritz potential energy method mostly concentrate near the center of the structure. Yet, a part of the master degrees of freedom are spread to both sides, which may be an important act to increase the higher order frequency.

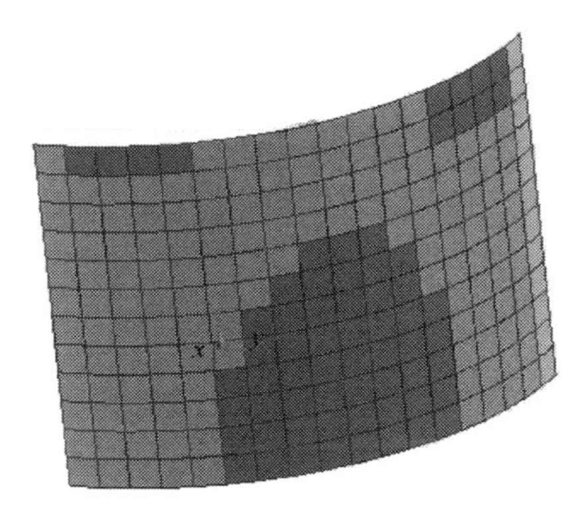

(a) Select 90 master degrees of freedom based on the Ritz vector method

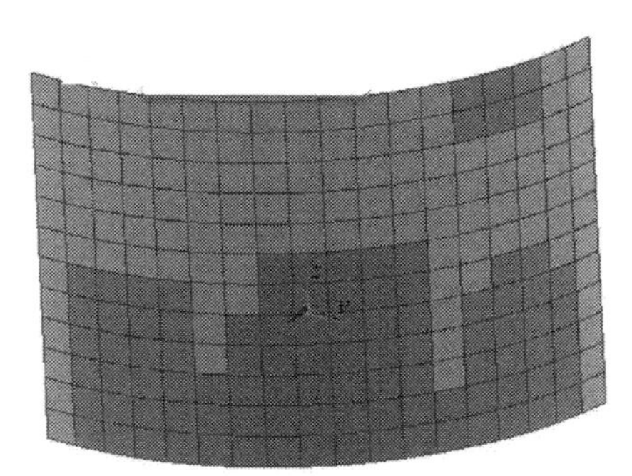

(b) Select 90 master degrees of freedom based on the nodal Leeds potential energy method

Figure 7-3 Cylindrical curved plate selected as the unit corresponding to the main degree of freedom

It can be seen from Figure 7-4 that under the same master degrees of freedom, the relative error of randomly selecting the master degrees of freedom is too large. Still, the error of selecting the master degrees of freedom based on the Ritz vector method is very small. In the first 30 order modes, the maximum relative error does not exceed 10%, while the minimum error of the master degrees of freedom is selected based on the nodal Ritz potential energy method. In the first 30 order modes, the maximum relative error is 3%. Table 7.1 is the natural frequency corresponding to the different calculation methods of the cylindrical curved plate, from which we can see the advantages of selecting the master degrees of freedom based on the nodal Ritz potential energy method.

Table 7.1 Natural frequencies corresponding to different calculation methods for cylindrical curved plates (400 master degrees of freedom) (Unit: Hz)

Modal	Finite element method	Randomly choose the main degree	Selecting the main degrees of freedom based on the Ritz vector method	Selection of main degrees of freedom based on nodal Leeds potential energy method
1	0.59448	0.59439	0.608297	0.594624
2	1.13384	1.338447	1.13791	1.134192
3	1.25966	1.506213	1.264697	1.260535
4	1.88371	2.268897	1.902101	1.888879
5	2.05532	3.999968	2.069269	2.057616
6	3.01821	4.850695	3.04936	3.034848
7	3.17274	5.279688	3.220343	3.204381
8	3.87476	5.398658	3.927064	3.900816
9	4.0099	6.861606	4.1114	4.036825
10	4.0583	7.205445	4.217862	4.110223
11	4.3671	7.263573	4.495803	4.405929
12	5.33462	8.256798	5.463962	5.367649
13	5.34684	8.478391	5.490919	5.420106
14	5.4016	8.576859	5.581511	5.475155
15	5.69404	9.204615	5.837194	5.73556
16	5.82552	9.715332	6.052567	5.86783
17	6.07753	10.07672	6.246831	6.150666
18	6.08887	10.60426	6.400604	6.192861
19	6.97324	11.05647	7.117503	7.026188
20	6.97969	11.37669	7.192666	7.089496
21	6.98439	11.59571	7.29423	7.136528

(Continued)

Modal	Finite element method	Randomly choose the main degree	Selecting the main degrees of freedom based on the Ritz vector method	Selection of main degrees of freedom based on nodal Leeds potential energy method
22	7.11592	12.17989	7.456346	7.181199
23	7.52813	13.35639	7.970673	7.712625
24	7.94012	13.66132	8.190527	8.07034
25	8.02193	13.79795	8.216696	8.097347
26	8.02921	14.10109	8.348083	8.148769
27	8.03269	14.64628	8.535974	8.203182
28	8.0455	15.4377	8.585916	8.259124
29	8.21062	15.87574	8.718633	8.29726
30	2253	15.95854	8.812292	8.377638

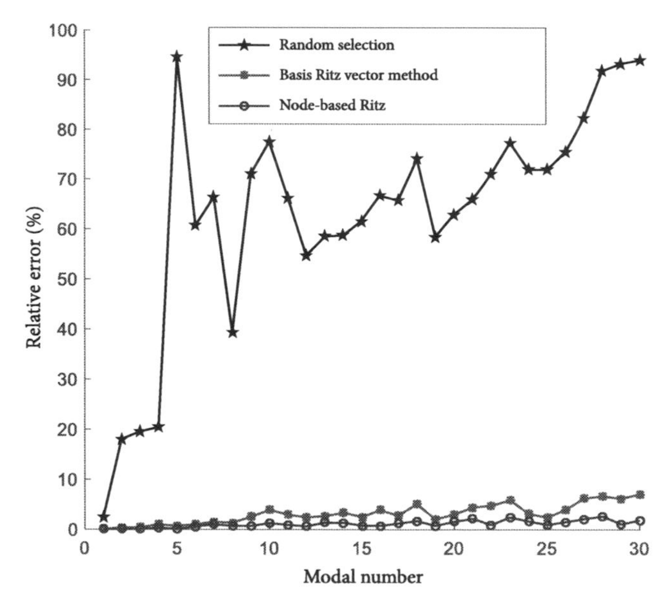

Figure 7-4 Comparison of the relative errors of the results obtained by selection method of different master degrees of freedom for cylindrical curved plate

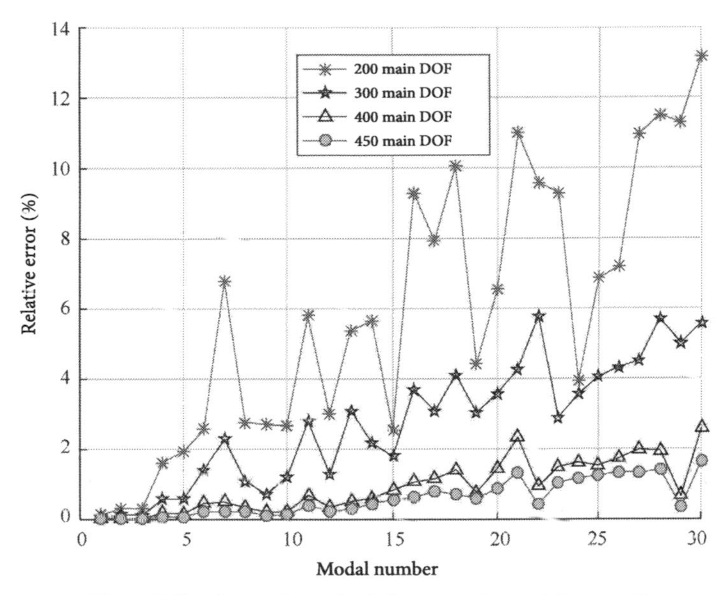

Figure 7-5 Comparison of relative error of calculation results

It can be seen from Figure 7-5 that when the master degrees of freedom are selected based on the node Ritz potential energy, when the number of master degrees of freedom increases, the relative error of the structural modes decreases. When there are 450 master degrees of freedom, the relative error of the first 30 order modes should not exceed 2% (at most). In addition, as the number of master degrees of freedom increases, the relative error of the structural modal number decreases further and becomes more slow. The relative error at 450 master degrees of freedom is about 1/3 of the relative error of all degrees of freedom. If you continue to increase the number of master degrees of freedom, although you can continue to reduce the relative error, it is of little significance for engineering applications. Therefore, as long as the relative error satisfies the accuracy, the method proposed in this chapter can be used.

7.4.2 Crankshaft

The diameter of the crankshaft is 6 mm, the diameter of the crank is 4 mm, the length of the crank is 4 mm, while the total length of the crank is 34 mm, and both ends are fixed as shown in Figure 7-6. The elastic modulus is 210 GPa, the Poisson's ratio is 0.3, and the density is 7850 kg/m^3. It uses SOLID92 elements for meshing, obtaining a total of 1022 elements, 2127 nodes, and 6231 degrees of freedom.

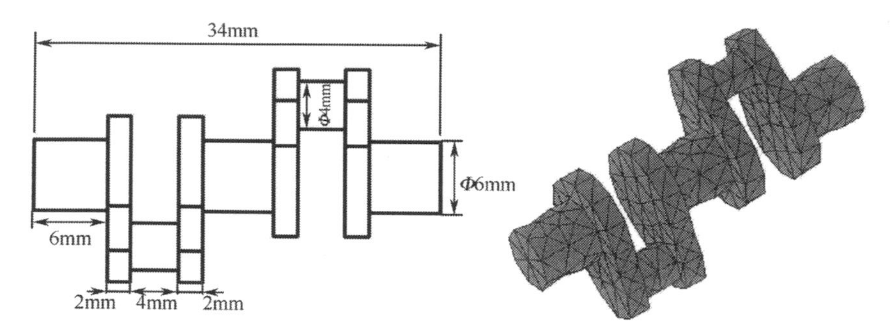

Figure 7-6 Crankshaft geometric

Figure 7-7 shows the corresponding units of the 500 master degrees of freedom selected by the below two methods (the darker cells). It can be seen that the master degrees of freedom selected based on the Ritz vector method are concentrated in the middle of the structure, and the master degrees of freedom are selected based on the nodal Ritz potential energy method that diffuses from the middle to both ends – this is very obvious.

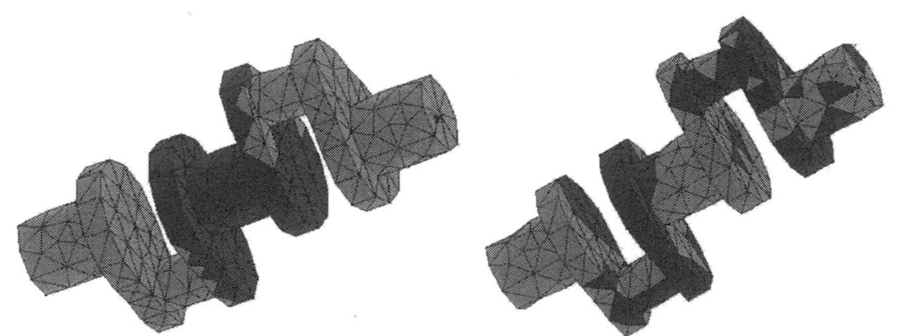

(a) 500 master degrees of freedom are selected based on the Ritz vector method

(b) 500 master degrees of freedom are selected based on the nodal Leeds potential energy method

Figure 7-7 Units corresponding to 500 main degrees of freedom selected by Element corresponding to 500 main degrees of freedom selected by two different methods

Figure 7-8 shows the selection of 1800 master degrees of freedom using random selection based on the Ritz vector method and the nodal Ritz potential energy method, and using the IRS method to construct the reduction system, and then solving the relative error of the obatined crankshaft modal. It can be seen from this that it is not feasible to randomly select the master degrees of freedom. The relative error of the results obtained by selecting the master degrees of freedom based on the node Ritz potential energy method is the smallest. In the first 30 order modes, the maximum relative error does not exceed 10%. The Ritz vector method selects the master degrees of freedom better than the random selection, and it has a maximum relative error of more than 20% among the first 30 orders of relative error.

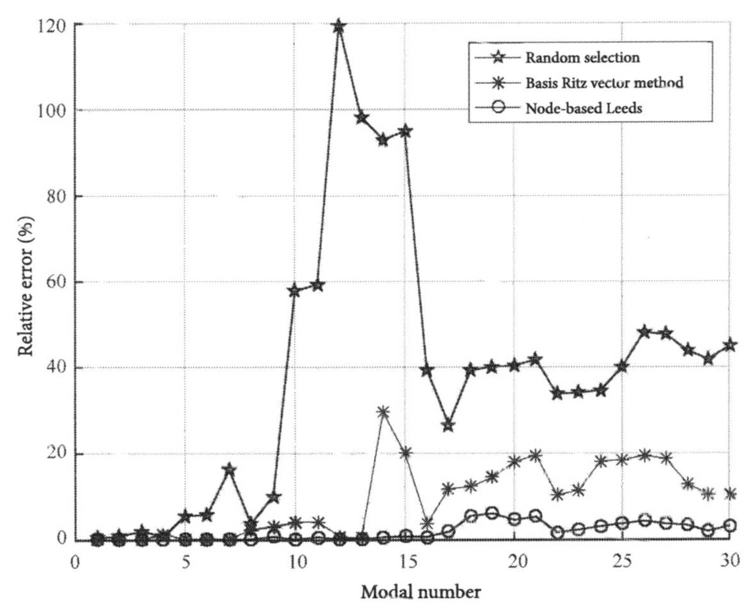

Figure 7-8 Comparison of the relative error of the results calculated by different methods of selecting the master degrees of freedom

Figure 7-9 shows the relative error comparison of the calculation results based on the nodal Ritz potential energy method to select the different numbers of master degrees of freedom. It can be seen from the figure that as the number of master degrees of freedom continues to increase, the relative error of the structural modal continues to decrease, but the magnitude of the decrease becomes smaller. When the number of master degrees of freedom is 2000 in the first 30 modes, the relative error of the structural mode does not exceed 5% at the most. Table 7.2 shows the natural frequencies corresponding to different calculation methods, from which we can see the advantages of choosing the main degree of freedom based on the nodal Ritz potential energy method.

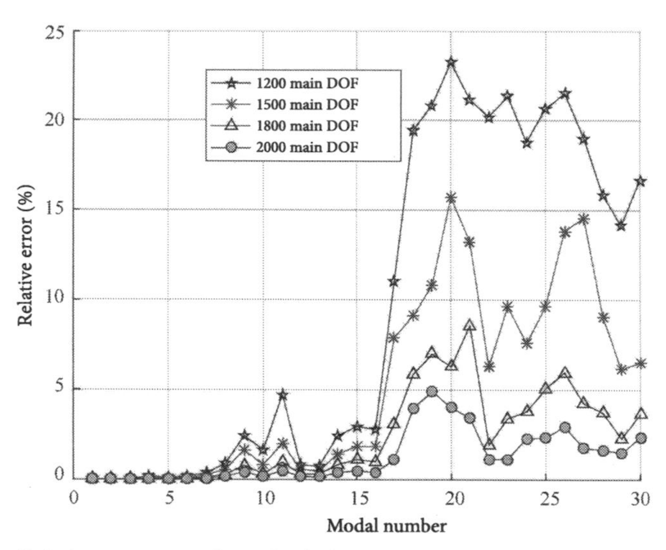

Figure 7-9 Relative error comparison of calculation results of selecting different numbers of master degrees of freedom based on the node Ritz potential energy method

Table 7.2 Natural frequencies corresponding to different calculation methods (1800 master degrees of freedom) (Unit: Hz)

Modal	Finite element method	Randomly select the main DOF	Selecting the main degrees of freedom based on the Ritz vector method	Selection of main degrees of freedom based on nodal Leeds potential energy method
1	9.6204621	9.66577909	9.628837946	9.620549
2	10.218049	10.3012917	10.21818102	10.21808
3	13.704415	13.97628862	13.70600513	13.70458
4	19.312367	19.45734276	19.60232902	19.3143
5	19.913589	21.02136738	19.93880641	19.91457
6	31.183477	32.9809545	31.27716915	31.18691
7	3.172743	41.0925088	35.42700862	35.38305
8	3.874757	42.20387261	41.63158519	40.77333
9	4.009897	45.21231666	42.23270455	41.09888
10	42.009326	66.2696177	43.68778218	42.07653
11	42.25162	67.36207042	43.86639523	42.3005
12	49.989035	109.7676844	50.27384058	50.03157

(Continued)

Modal	Finite element method	Randomly select the main DOF	Selecting the main degrees of freedom based on the Ritz vector method	Selection of main degrees of freedom based on nodal Leeds potential energy method
13	56.313051	111.6456385	56.55015089	56.35574
14	61.005197	117.7396863	79.08624849	61.17104
15	66.30236	129.2743599	81.7511321	66.47433
16	95.954883	133.79121	99.7605438	96.21729
17	109.61362	138.8162823	122.4226632	110.8421
18	109.73629	152.7806326	123.1270051	111.5437
19	109.9529	154.1525155	126.1898216	115.298
20	111.93817	157.197504	131.8098396	115.5989
21	115.6522	164.0177176	137.5171505	116.2616
22	126.70962	169.4128793	139.690564	128.0151
23	127.36071	170.6447539	141.7406596	128.1126
24	131.76773	177.2383768	155.2585013	134.3963
25	132.7325	185.9291271	156.7316567	134.9403
26	133.11654	197.2350036	158.9077245	135.9264
27	137.195	202.6391901	162.1654423	138.0203
28	146.39732	210.5770249	165.5240438	147.8184
29	151.28143	214.7553653	167.2064854	153.4004
30	152.62456	221.0713768	168.3372615	154.3122

7.5 Summary of this chapter

In this chapter, the master degrees of freedom are selected by the Ritz vector method and the node Ritz potential energy method, and the reduced system is constructed by the IRS method. Finally, the generalized Schur decomposition method is used to solve the reduced structural dynamic equation. In addition, through the analysis of two examples of cylindrical curved plate and crankshaft, the effectiveness of the method of selecting the main degree of freedom based on the nodal Ritz potential energy method is illustrated, and the following conclusions are obtained:

1) By converting the modal space to the Ritz vector space, it is possible to capture very precise dynamic characteristics with a few Ritz vectors. Due to its excessive

emphasis on low-order frequencies, the accuracy of high-order frequencies needs to be improved by further defining weighting coefficients.

2) The node Ritz potential energy method can capture a more suitable main degree of freedom than the Ritz vector method, which can better reflect the dynamic characteristics of the structure. Under the same conditions, it can obtain higher accuracy and can be used for the next steps of laying the foundation for efficient dynamic response optimization.
3) In the process of structural reduction, the number of master degrees of freedom is generally selected as 1/3 of the total number of degrees of freedom. For example, the total number of degrees of freedom of the cylindrical crank plate is 1326, and the calculation of 450 master degrees of freedom can obtain the accuracy of the relative error of the first 30 modes is not more than 2%. The total number of degrees of freedom of the crankshaft is 6231, and the calculation of 2000 master degrees of freedom show that the accuracy of the relative error of the first 30 modes is not more than 5%.

References

[1] Fu M C. Optimization via simulation: A review [J]. Annals of Operations Research, 1994, 53 (1): 199–247.

[2] Boesel J, Bowden R O, Glover F, et al. Future of simulation optimization [C] // Simulation Conference, 2001. Proceedings of the Winter. IEEE, 2001: 1466–1469.

[3] Azadivar F. Simulation optimization methodologies [C]. Simulation Conference Proceedings. IEEE, 1999: 93–100.

[4] Tran D M. Component mode synthesis methods using partial interface modes: Application to tuned and mistuned structures with cyclic symmetry [J]. Computers & Structures, 2009, 87 (17): 1141–1153.

[5] Craig R R. A Brief Tutorial on Substructure Analysis and Testing [C]. Proceedings of the 18th IMAC Conference on Computational Challenges in Structural Dynamics 2000; 1 (2): 899–908.

[6] Bouhaddi N, Lombard J P. Improved free-interface substructures representation method [J]. Computers & Structures, 2000, 77 (3): 269–283.

[7] Tran D-M. Méthodes de synthèse modale mixtes [J]. Rev Eur Eléments Finis, 1992; 1 (2): 137–179.

[8] Benfield W, Hruda R F. Vibration Analysis of Structures by Component Mode Substitution [J]. Aiaa Journal, 1971, 9 (7): 1255–1261.

[9] Hurty W C. Vibrations of structural systems by component mode synthesis [J]. Transactions of the American Society of Civil Engineers, 1960, 126 (1): 157–175.

[10] Yan Y J, Cui P L, Hao H N. Vibration mechanism of a mistuned bladed-disk [J]. Journal of Sound & Vibration, 2008, 317 (1-2): 294–307.

[11] Shanmugam A, Padmanabhan C. A fixed–free interface component mode synthesis method for rotordynamic analysis [J]. Journal of Sound & Vibration, 2006, 297 (3): 664–679.

[12] Zhang Yan, Ma Feng, Song Jun, Ren Chao. Application of substructure method in the model of complicated frequency response for rapid analysis [J]. Automobile Technology, 2017 (08): 59–62.

[13] Gao Pengfei, Mao Huping, Yang Yuguang, Zhang Qiang, Zhu Qin. Temperature distribution and mechanical stress analysis of piston based on substructure method [J]. Machine Design and Research, 2017, 33 (04): 163–166.

[14] Gao Pengfei, Mao Huping, Yang Yuguang, Zhang Qiang. Application of substructure method in optimization of piston structure [J]. Modular Machine Tool & Automatic Manufacturing Technique, 2017 (03): 10–13.

[15] Liu Bo, Dong Xiao-rui, Pan Cui-li, Mao Hu-ping, Zhang Yan-gang. Parametric design method for optimizing local features [J]. Manufacturing Automation, 2014, 36 (13): 132–135.

[16] Li Yun. Application of substructure in large chassis tower connecting structure [D]. Xi' An University of Architecture And Technology, 2016.

[17] Chai Guodong, Wang Shunhui, Chai Guoqiang. The electronic equipment modal analysis based on the sub-stucture [J]. Journal of Mechanical Strength, 2015, 37 (04): 790–792.

[18] Zhang Mingming, Zhao Jianhua, Zhang Ruibo. Research on the finite element modeling method of the crankshaft in diesel engine based on substructure [J]. Small Internal Combustion Engine And Vehicle Technique, 2015, 44 (04): 27–31, 96.

[19] Ding Yang, Li Hao, Shi Yanchao, Li Zhongxian. Rapid evaluation method of progressive collapse resistant capacity for steel frame structures based on sub-structure [J]. Journal of Building Structures, 2014, 35 (06): 109–114.

[20] Ding Xiaohong, Zhao Xinfang, Wang Haihua, Xu Feng. A study on the component topology optimization with successive approximation based on substructure technique [J]. Automotive Engineering, 2014, 36 (05): 639–642.

[21] Zhang Sheng, Bai Yang, Yin Jin, Chen Biaosong. Comparative study on multiple multi-level substructure method and modal synthesis method [J]. Applied Mathematics and Mechanics, 2013, 34 (02): 118–126.

[22] Zhang Fan, Gang Xianyue, Chai Shan, Pang Jianwu. Topological optimization of bus under multi-conditions based on equivalent load and substructure method [J]. Journal of Machine Design, 2013, 30 (03): 62–67.

[23] Li Zhigang, Chu Yuchuan, Zheng Feng. Finite element analysis of elevated railway floating bridge based on substructure method [J]. Journal of PLA University of Science and Technology (Natural Science Edition), 2013, 14 (03): 277–282.

[24] Mao Huping, Gao Pengfei, Qin Jianjian. Continuous structural optimzation based on substructure method with local characteristic [J]. Computer Integrated Manufacturing Systems, 2018, 24 (08): 2079–2087.

[25] Mao Huping, Wu Yizhong, Chen Liping. SQP parallel optimization algorithm based on multi-domain simulation [J]. China Mechanical Engineering, 2009, 20 (15): 1823–1829.

[26] Jeong J, Baek S, Cho M. Dynamic condensation in a damped system through rational selection of primary degrees of freedom [J]. Journal of Sound & Vibration, 2012, 331 (7): 1655–1668.

[27] Kim K O, Choi Y J. Energy Method for Selection of Degrees of Freedom in Condensation [J]. Aiaa Journal, 2000, 38 (7): 1253–1259.

[28] Cho M, Kim H. Element-Based Node Selection Method for Reduction of Eigenvalue Problem [M] // Mathematical and Numerical Aspects of Wave Propagation WAVES 2003. Springer Berlin Heidelberg, 2003: 1677–1684.

[29] Zienkiewicz O C, Campbell J S. Optimum structural design [M]. John Wiley & Sons, 1973.

[30] Barthelemy B, Chen C T, Haftka R T. Sensitivity approximation of the static structural response. In First World Congress on Computational Mechanics, Austin, TX, Sept. 1986.

[31] Pauli Pedersen, Gengdong Cheng, John Rasmussen. On Accuracy Problems for Semi-Analytical Sensitivity Analyses [J]. Mechanics of Structures & Machines, 1989, 17 (3): 373–384.

[32] Olhoff N, Rasmussen J. Study of inaccuracy in semi-analytical sensitivity analysis — a model problem [J]. Structural & Multidisciplinary Optimization, 1991, 3 (4): 203–213.

[33] Cheng G, Olhoff N. Rigid body motion test against error in semi-analytical sensitivity analysis [J]. Computers & Structures, 1993, 46 (3): 515–527.

[34] Boer H D, Keulen F V. Refined semi-analytical design sensitivities [J]. International Journal of Solids & Structures, 2000, 37 (46–47): 6961–6980.

[35] Oral S. A Mindlin plate finite element with semi-analytical shape design sensitivities [J]. Computers & Structures, 2000, 78 (1): 467–472.

[36] Cho M, Kim H. A refined semi-analytic design sensitivity based on mode decomposition and Neumann series [J]. International Journal for Numerical Methods in Engineering, 2004, 62 (62): 19–49.

[37] Kang B S, Park G J, Arora J S. A review of optimization of structures subjected to transient loads [J]. Structural & Multidisciplinary Optimization, 2006, 31 (31): 81–95.

[38] Zhang Yangang, Su Tiexiong, Mao Huping, Guo Zhiming, Wang Zhibin, Li Jianjun. Critical time points identification method for solution space of dynamic stress based on spectral element [J]. Journal of Mechanical Engineering, 2014, 50 (5): 82–84.

[39] Zhang Yangang, Su Tiexiong, Mao Huping, Guo Zhiming. Transformation method of equivalent static loads from dynamic loads based on structural potential principle [J]. Transactions of Beijing Institute of Technology, 2014, 34 (5): 454–459.

[40] Mao Huping, Dong Xiaorui, Guo Baoquan, and Wang Qiang. Structure dynamic response optimization under equivalent static loads of each node based on mode superposition [J]. Journal of Computer-Aided Design & Computer Graphics, 2017, 29 (09): 1759–1766.

[41] Zhang Yangang, Mao Huping, Su Tiexiong, Li Kun, Wang Jun. Critical time identification for global dynamic stress based on spectral element interpolation of solution space [J]. China Mechanical Engineering, 2016, 27 (05): 688–693.

[42] Patera A T. A spectral element method for fluid dynamics; laminar flow in a channel expansion [J]. Journal of Computational Physics, 1984, 54 (3): 468–488.

[43] Bueno-Orovio A, Pérez-García V M. Spectral smoothed boundary methods: The role of external boundary conditions [J]. Numerical Methods for Partial Differential Equations, 2010, 22 (2): 435–448.

[44] Kurdi M H, Beran P S. Spectral element method in time for rapidly actuated systems [J]. Journal of Computational Physics, 2008, 227 (3): 1809–1835.

[45] Mao Huping, Wu Yizhong, Chen Liping. Dynamic response optimization based on temporal spectral element method [J]. Journal of Mechanical Engineering, 2010, 46 (16): 79–87.

[46] Henderson R D, Karniadakis G E. Unstructured spectral element methods for simulation of turbulent flows [M]. Academic Press Professional, Inc. 1995.

[47] Pathria D, Karniadakis G E. Spectral Element Methods for Elliptic Problems in Nonsmooth Domains [J]. Journal of Computational Physics, 1995, 122 (1): 83–95.

[48] Hesthaven J S, Gottlieb D. A Stable Penalty Method for the Compressible Navier-Stokes Equations. I. Open Boundary Conditions [J]. Siam Journal on Scientific Computing, 2015, 20 (1): 62–93.

[49] Priolo E, Seriani G. A numerical investigation of Chebyshev spectral element method for acoustic wave propagation [C]. Proc. Imacs Conf. on Comp. Appl. Math. 1991: 551–556.

[50] Mao Huping, Qiao Wenyuan, Guo Baoquan, Wang Qiang, Dong Xiaorui. Application of sem based on accumulation elements to dynamic analysis of the structures subjected to impact loads [J]. Noise and Vibration Control, 2016, 36 (6): 45–50.

[51] Mao Huping, Liu Xiaojie, You Guodong, Zhang Yangang, Dong Xiaorui. Chebyshev spectral element method for solving vibration problems under arbitrary load [J]. Journal of Machine Design, 2017 (10): 49–55.

[52] Mao Huping, Wang Weineng, Xu Yan-fang, Zhang Yangang, Dong Xiaorui. Chebyshev spectral element method for analysis of nonlinear vibration problems [J]. Noise and Vibration Control, 2015, 35 (1): 73–77.

[53] Mao Huping, Su Tiexiong, Li Jianjun. A step-by-step temporal spectral element method for system dynamic response simulation [J]. Journal of North University of China (Natural Science Edition), 2013 (4): 424–430.

[54] Mao Huping, Su Tiexiong, Li Jianjun. The dynamic optimization based on combination of adaptive multi-metamodeling and temporal spectral element method [J]. Journal of Computer-Aided Design & Computer Graphics, 2013, 25 (11): 1725–1734.

[55] Thomas J. R. Hughes. The Finite Element Method—Linear Static and Dynamic Finite Element Analysis [M], Prentice-Hall, lnc, Englewood Cliffs, New Jersey, 1987.

[56] Niordson F I. On the optimal design of a vibrating beam (Supported beam analysis for finding best possible tapering optimizing highest natural frequency for lowest mode of lateral vibration [J]. Quarterly of Applied Mathematics, 1965, 23: 47–53.

[57] Cassis J H, Schmit L A. Optimum structural design with dynamic constraints [J]. Journal of the Structural Division, 1976, 102 (10): 2053–2071.

[58] Wang D, Zhang W H, Jiang J S. Truss optimization on shape and sizing with frequency constraints [J]. Aiaa Journal, 2004, 42 (3): 622–630.

[59] Qin Jianjian, Mao Huping. Diesel engine connecting rod structure optimization design based on miga algorithm [J]. Machinery Design & Manufacture, 2017 (04): 218–221.

[60] Lin J H, Che W Y, Yu Y S. Structural optimization on geometrical configuration and element sizing with statical and dynamical constraints [J]. Computers & Structures, 1982, 15 (5): 507–515.

[61] Pantelides C P, Tzan S R. Optimal design of dynamically constrained structures [J]. Computers & structures, 1997, 62 (1): 141–149.

[62] Min S, Kikuchi N, Park Y C, et al. Optimal topology design of structures under dynamic loads [J]. Structural optimization, 1999, 17 (2–3): 208–218.

[63] Du J, Olhoff N. Minimization of sound radiation from vibrating bi-material structures using topology optimization [J]. Structural and Multidisciplinary Optimization, 2007, 33 (4–5): 305–321.

[64] Mao Huping, Wu Yizhong, Li Jianjun, Wang Yinghong. Temporal spectral element method and its application in optimization of dynamic response [J]. Journal of Vibration Engineering, 2013 (03): 395–403.

[65] Mao Huping. Dynamic response optimization method based on simulation model [M], Beijing: Publishing House of Electronics Industry, 2014, 3.

[66] Gu L. A comparison of polynomial based regression models in vehicle safety analysis [C]. Proc. 2001 ASME design engineering technical conferences-design automation conference, Pittsburgh, 2001: 196–121.

[67] Zou Linjun, Wu Yizhong, Mao Huping. Incremental Kriging Model Rebuilding Method and its Application in Efficient Global Optimization [J]. Journal of Computer-Aided Design & Computer Graphics, 2011, 23 (04): 649–655.

[68] Mao Huping, Wu Yizhong, Chen Liping. Multivariate adaptive regression splines based simulation optimization using move-limit strategy [J]. Journal of Shanghai University (English Edition), 2011, 15 (06): 542–547.

[69] Jin R, Chen W, Sudjianto A. An efficient algorithm for constructing optimal design of computer experiments [J]. Journal of Statistical Planning & Inference, 2003, 134 (1): 268–287.

[70] Xu Yueji. Vibration analysis and diagnosis on gear failure of gear box [J]. Journal of Machine Design, 2009, 26 (12): 68–71.

[71] Weaver Jr W, Timoshenko S P, Young D H. Vibration problems in engineering [M]. John Wiley & Sons, 1990.

[72] Lin Weijun. A Chebyshev spectral element method for elastic wave modeling [J]. Acta Acustica, 2007, 32 (6): 525–533.

[73] Zhang Jin, Wang Guoping, Rui Xiaoting. Vibration analysis of systems with random parameters using perturbation transfer matrix method [J]. Journal of Machine Design, 2015 (10).

[74] Zhang Feipeng, Guo Junyi, Huang Cheng. Some studies on the Chebyshev collocation method for solving ordinary differential equations [J]. Annals of Shanghai Observatory Academia Sinica, 1998 (19): 6–15.

[75] Li Shuping. Barycentric interpolation collocation method for numerical analysis of mechanical vibrations [D]. Shandong University, 2007.

[76] Wang Zhaoqing, Li Shuping, Tang Bingtao, Zhao Xiaowei. High precision numerical analysis of vibration problems under pulse excitation [J]. Journal of Mechanical Engineering, 2009, 45 (1): 288–292.

[77] Orszag S A. Numerical methods for the simulation of turbulence [J]. Physics of Fluids (1958–1988), 2004, 12 (12): II-250-II-257.

[78] Guo B. Spectral methods and their applications [M]. World Scientific, 1998.

[79] Canuto C, Hussaini M Y, Quarteroni A, et al. Spectral methods: evolution to complex geometries and applications to fluid dynamics [M]. Springer, 2007.

[80] [80] Komatitsch D, Tromp J. Introduction to the spectral element method for three-dimensional seismic wave propagation [J]. Geophysical Journal International, 1999, 139 (3): 806–822.

[81] Deville M O, Fischer P T, Mund E H. High-order methods for incompressible fluid flow [M]. Cambridge University Press, 2002.

[82] Zhu W, Kopriva D A. A spectral element approximation to price European options. II. The Black-Scholes model with two underlying assets [J]. Journal of Scientific Computing, 2009, 39 (3): 323–339.

[83] Taylor M, Tribbia J, Iskandarani M. The spectral element method for the shallow water equations on the sphere[J]. Journal of Computational Physics, 1997, 130 (1): 92–108.

[84] Bar-Yoseph P Z, Fisher D, Gottlieb O. Spectral element methods for nonlinear temporal dynamical systems [J]. Computational mechanics, 1996, 18 (4): 302–313.

[85] Zrahia U, Bar-Yoseph P. Space-time spectral element method for solution of second-order hyperbolic equations [J]. Computer Methods in Applied Mechanics and Engineering, 1994, 116 (1): 135–146.

[86] Pozrikidis C. Introduction to Finite and Spectral Element Methods Using Matlab, Chapman and Hall/CRC, 2005.

[87] Parter S V. On the Legendre-Gauss-Lobatto Points and Weights [J]. Journal of Scientific Computing, 1999, 14 (4): 347–355.

[88] Zhao J M, Liu L H. Solution of radiative heat transfer in graded index media by least square spectral element method [J]. International Journal of Heat and Mass Transfer, 2007, 50 (13): 2634–2642.

[89] Xiao Feng, Chen Yong, Ma Chao, Sun Jingya, Hua Hongxin. Dynamic response and shock resistance of chiral honeycomb rubber claddings subjected to underwater explosion [J]. Noise And Vibration Control, 2013 (04): 44–49.

[90] Wang Gongxian, Shen Rongying. Study on dynamic properties of the crane jib under lifting shock load [J]. Journal of Mechanical Strength, 2005, 27 (5): 561–566.

[91] Tang Jinyuan, Peng Fangjin, Huang Yunfei. Numerical analysis of dynamic stress variation in spur gear under impact loads [J]. Journal of Vibration and Shock, 2009, 28 (08): 138–143.

[92] Li Yongqiang, Huan Qiang. Dynamic response of honeycomb sandwich panel under impact loading based on ansys [J]. Journal of Northeastern University (Natural Science), 2015, 36 (6): 858–862.

[93] Kurdi M H, Beran P S. Optimization of dynamic response using a monolithic-time formulation [J]. Structural and Multidisciplinary Optimization, 2009, 39 (1): 83–104.

[94] Wen Bangchun, Li Yinong, Xu Peimin, etc. Engineering nonlinear vibration [M]. Science Publishing Company, 2007.

[95] Orszag S A. Numerical Methods for the Simulation of Turbulence [J]. Physics of Fluids, 1969, 12 (12): II-250-II-257.

[96] Guo B. Spectral methods and their applications [M]. World Scientific, 1998.

[97] Boyd J P. Chebyshev and Fourier spectral methods [M]. Courier Dover Publications, 2013.

[98] Valenciano J, Chaplain M A J. A laguerre-legendre spectral-element method for the solution of partial differential equations on infinite domains: Application to the diffusion of tumour angiogenesis factors [J]. Mathematical and computer modelling, 2005, 41 (10): 1171–1192.

[99] High-order methods for incompressible fluid flow [M]. Cambridge University Press, 2002.

[100] Zhu W, Kopriva D A. A spectral element approximation to price European options with one asset and stochastic volatility [J]. Journal of Scientific Computing, 2010, 42 (3): 426–446.

[101] Zhu W, Kopriva D A. A spectral element approximation to price European options. II. The Black-Scholes model with two underlying assets [J]. Journal of Scientific Computing, 2009, 39 (3): 323–339.

[102] Li Fucai, Peng Haikuo, Sun Xuewei, Wang Jinfu, Meng Guang. Guided wave propagation mechanism and damage detection in plate structures using spectral element method [J]. Journal of Mechanical Engineering, 2013, 48 (21): 57–66.

[103] Zhao J M, Liu L H. Least-squares spectral element method for radiative heat transfer in semitransparent media [J]. Numerical Heat Transfer, Part B: Fundamentals, 2006, 50 (5): 473–489.

[104] Lin Weijun. A Chebyshev spectral element method for elastic wave modeling [J]. Acta Acustica, 2007, 32 (6): 525–533.

[105] Geng Yanhui, Qin Guoliang, Wang Yang, He Wei. The research of space-time coupled spectral element method for acoustic wave equation [J]. Acta Acustica, 2013, 38 (3): 306–318.

[106] Bar-Yoseph P, Moses E, Zrahia U, et al. Space-temporal spectral element methods for one-dimensional nonlinear advection-diffusion problems [J]. Journal of Computational Physics, 1995, 119 (1): 62–74.

[107] Zrahia U, Bar-Yoseph P. Space-time spectral element method for solution of second-order hyperbolic equations [J]. Computer Methods in Applied Mechanics and Engineering, 1994, 116 (1-4): 135-146.

[108] Bar-Yoseph P Z, Fisher D, Gottlieb O. Spectral element methods for nonlinear spatio-temporal dynamics of an Euler-Bernoulli beam [J]. Computational Mechanics, 1996, 19 (1): 136-151.

[109] Taniguchi. Translated by Yin Chuanjia et al. Vibration Engineering: vol. 2 [M]. Press of Engineering Industry, 1986.

[110] Lu Zhongrong, Liu Jike. Vibration analysis of pendulum [J]. Journal of Jinan University (Natural Science), 1999, 20 (1): 42-45.

[111] Zhou Kaihong, Wang Yuanxun, Li Chunzhi. The application of differential quadrature method in nonlinear vibration analysis of simple pendulu [J]. Mechanics in Engineering, 2003, 25 (3): 50-52.

[112] Wang Q, Arora J S. Alternative Formulations for Transient Dynamic Response Optimization [J]. Aiaa Journal, 2005, 43 (43): 2188-2195.

[113] Choi D H, Park H S, Kim M S. A direct treatment of min-max dynamic response optimization problems. AIAA-93-1352-CP.

[114] Park S, Kapania R K, Kim S J. Nonlinear Transient Response and Second-Order Sensitivity Using Time Finite Element Method [J]. Aiaa Journal, 2012, 37 (5): 613-622.

[115] Choi W S, Park G J. Structural optimization using equivalent static loads at all time intervals [J]. Computer Methods in Applied Mechanics & Engineering, 2002, 191 (19): 2105-2122.

[116] Grandhi R V, Haftka R T, Watson L T. Design-oriented identification of critical times in transient response [J]. Aiaa Journal, 1986, 24 (4): 649-656.

[117] Kurdi M, Beran P. Optimization of Dynamic Response Using Temporal Spectral Element Method [C]. Aiaa Aerospace Sciences Meeting and Exhibit, 2008: 83-104.

[118] Hang E J, Arora J S, Applied optimal design: mechanical and structural systems [M], New York, Wiley, 1979.

[119] Xue Dingyu, Chen Yangquan. The solution of higher applied mathematics problems with MATLAB [M]. Beijing: Tsinghua University Press, 2004.

[120] Simpson T W, Poplinski J D, Koch P N, et al. Metamodels for Computer-based Engineering Design: Survey and recommendations [J]. Engineering with Computers, 2001, 17 (2): 129-150.

[121] Hsieh C C, Arora J S. Design sensitivity analysis and optimization of dynamic response [J]. Computer Methods in Applied Mechanics & Engineering, 1984, 43 (2): 195-219.

[122] Hsieh C C and Arora J S. A Hybrid Formulation for Treatment of Point-Wise State Variable Constraints in Dynamic Response Optimization [J]. Computer Methods in Applied Mechanics and Engineering, 1985, 48: 171-189.

[123] Paeng J K, Arora J S. Dynamic Response Optimization of Mechanical Systems with Multiplier Methods [J]. Journal of Mechanical Design, 1989, 111 (1): 73-80.

[124] Haftka R T, Gürdal Z. Elements of Structural Optimization [M]. Springer Science & Business Media, 1991.

[125] Kang B S, Choi W S, Park G J. Structural optimization under equivalent static loads transformed from dynamic loads based on displacement [J]. Computers & Structures, 2001, 79 (2): 145-154.

[126] Park G J. Analytic methods for design practice [M]. Springer Science & Business Media, 2007.

[127] Kang B S, Park G J, Arora J S. A review of optimization of structures subjected to transient loads [J]. Structural and Multidisciplinary Optimization, 2006, 31 (2): 81-95.

[128] Park K J, Lee J N, Park G J. Structural shape optimization using equivalent static loads transformed from dynamic loads [J]. International journal for numerical methods in engineering, 2005, 63 (4): 589–602.

[129] Dai Jianglu. The Design Optimization of Vehicle Frontal Crashworthiness Based on ESLMG and Target Curve[D]. Changsha: Hunan University, 2016.

[130] He Xinfeng. The fatigue design of the auxiliary frame of the agitator truck based on equivalent static loads and the response surface method [D]. Changsha: Hunan University, 2012.

[131] Lu Shanbin, Jiang Weibo, Zuo Wenjie. Size and morphology crashworthiness optimization for automotive frontal structures using equivalent static loads method[J]. Journal of Vibration and Shock, 2018 (7): 56–61.

[132] Gao Yunkai, Tian Linli. Topology optimization of automotive body crashworthiness design with equivalent static loads method [J]. Journal of Tongji University (Natural Science), 2017 (3): 87–93.

[133] Huang Yuhan, Yang Zhijun, Cai Tiegen, et al. Structural dynamic optimization of high-speed light load robot based on equivalent static load method (ESLM) [J]. Tool Engineering, 2016, 50 (4): 27–31.

[134] Zhang Heng, Ding Xiaohong, Duan Pengyun. Structural optimization of machine tool slide based on equivalent static load method [J]. Journal of Machine Design, 2016, 33 (1): 55–59.

[135] Huang Wulong. The lightweight design of large complex structures based on equivalent static load method [D]. Guangdong University of Technology, 2013.

[136] Duan Pengyun, Ding Xiaohong. Structural design optimization method for moving parts of machine tool based on equivalent static load theory [J]. Journal of University of Shanghai for Science and Techology, 2015, 37 (6): 583–588.

[137] Rui Qiang, Wang Hongyan, Tian Honggang. Structural dynamic optimization based on equivalent static load method [J]. Automotive Engineering, 2014, 36 (1): 61–65.

[138] Li Ming, Tang Wencheng. Reliability-based topology optimization of internal parameter structures using equivalent static loads [J]. Chinese Journal of Computational Mechanics, 2014, 31 (3): 297–302.

[139] Zhang Yangang, Mao Huping, Su Tiexiong, etc. Method of equivalent static loads based on energy principle and its application in dynamic optimization design of diesel engine piston [J]. Journal of Shanghai Jiao Tong University, 2015, 49 (9): 1293–1299.

[140] Zhang Yangang, Mao Huping, Su Tiexiong, etc. Energy equivalent static load method based on critical time point and structure dynamic response optimization [J]. Journal of Mechanical Engineering, 2016, 52 (9): 151–157.

[141] Chen Kechang. Substructure method [J]. Central South Highway Engineering, 1980, (02): 45–52.

[142] Lou Menglin. Static substructure method for dynamic analysis of structures [J]. Engineering Mechanics, 1990, 7 (1): 57–66.

[143] Zheng Xinyuan. Substructure method in finite element [J]. Thermal Turbine, 1991 (1): 50–58.

[144] Li Yuanke, Hu Yujin. Substructure method for the 3D finite element analysis of rolling bearings[J]. Bearing, 1992 (3): 2-7.

[145] Zhao Wenzhong, Gu Qiannong, Yu Lianyou. Substructure method of displacement sensitivity analysis [J]. Journal of Dalian Railway Institute, 1996 (2): 4–8.

[146] Du Jiazheng, Xiao hui, Namho. Topology optimization for continuous structures with internal force constraints based on the substructure [J]. Journal of Beijing University of Technology, 2016, 42 (12): 1818–1821.

[147] Sui Y, Du J, Guo Y. Independent continuous mapping for topological optimization of frame structures [J]. Acta Mechanica Sinica, 2006, 22 (6): 611–619.

[148] Shu Lei, Fang Zongde, Dong Jun, et al. Design of composite domain topology optimization for vehicle substructure [J]. Automotive Engineering, 2008, 30 (5): 444–448.

[149] Jiang Haoran, Wang Tao, Zhou Huimeng, et al. A method to coordinate substructures based on interface elements [J]. Engineering Mechanics, 2017, 34 (1): 171–179.

[150] Aminpour M, Ransom J, McCleary S. Coupled analysis of independently modeled finite element subdomains [C] // 33rd Structures, Structural Dynamics and Materials Conference. 1992: 2235.

[151] Wang Bo, Sun Wei, Wen Bangchun. Inherent characteristic solution of motorized spindle based on receptance coupling substructure analysis [J]. Computer Integrated Manufacturing Systems, 2012, 18 (2): 422–426.

[152] Zhang Bao, Sun Qin. A substructure sensitivity method for large-scale structure [J]. Advances in Aeronautical Science and Engineering, 2014 (4): 475–480.

[153] Fonseka M C M. A sub-structure condensation technique in finite element analysis for the optimal use of computer memory [J]. Computers & structures, 1993, 49 (3): 537–543.

[154] Zhang Fan, Gang Xianyue, Chai Shan, etc. Topological optimization of bus under multi-conditions based on equivalent load and substructure method [J]. Journal of Machine Design, 2013, 30 (3): 62–67.

[155] Zhang Zaofa, Zhu Zhuangrui, Sun Lingyu, et al. An equivalent substructure method for dynamic sensitivity stress analysis of the white body [J]. Journal of Southeast University (Natural Science Edition), 2001, 31 (2): 39–41.

[156] Wang Fei, Jiang Nan. 3D dynamic analysis of soil-structure interaction system based on mixed linear-nonlinear substructure method [J]. Engineering Mechanics, 2012, 29 (1): 155–161.

[157] Gu Yuanxian, Zhang Hongwu, Liu Shutian, et al. The methods and application of structural layout and dynamic design optimization [J]. Bulletin of National Natural Science Foundation of China, 1998, 12 (4): 276–278.

[158] Zhao Ning, Wu Liyan, Liu Geng, et al. Shape optimization design for dynamic characteristics of blade disk structure [J]. Gas Turbine Experiment and Research, 1999 (3): 45–47.

[159] Zhang Junnuo, Wang Ruilin, Li Yongjian, et al. Optimization research on the dynamic characteristic of machine gun structure [J]. Journal of Test And Measurement Technology, 2007, 21 (3): 251–255.

[160] Li Xiaogang, Cheng Jin, Liu Zhenyu, et al. Robust optimization for dynamic characteristics of mechanical structures based on double renewal kriging model [J]. Journal of Mechanical Engineering, 2014, 50 (3): 165–173.

[161] Akl W, El-Sabbagh A, Baz A. Optimization of the static and dynamic characteristics of plates with isogrid stiffeners [J]. Finite Elements in Analysis & Design, 2008, 44 (8): 513–523.

[162] Feng H, Zhan Y, Wang X. Dynamic Characteristics Analysis and Structure Optimization Study of Glaze Spraying Manipulator [C] // MATEC Web of Conferences. EDP Sciences, 2016, 70: 02005.

[163] Sun Guozheng, Zhang Qing. Sensitivity analysis of generalized geometric programming and its usage in optimal design of crane structure [J]. Journal of Computer-Aided Design & Computer Graphics, 1993 (4): 297–305.

[164] Oguamanam D C D, Liu Z S, Hansen J S. Natural Frequency Sensitivity Analysis with Respect to Lumped Mass Location [J]. Aiaa Journal, 1971, 37 (8): 928–932.

[165] Liu Yuan, Liu Xuefeng, Deng Jiansong, et al. Global structural optimization of 3D models based on modal analysis [J]. Journal of Computer-Aided Design & Computer Graphics, 2015 (4): 590–596.

[166] Su Xindong, Guan Dihua. Reduction of brake squeal using substructure dynamic characteristics optimization [J]. Automotive Engineering, 2003, 25 (2): 167–170.

[167] Yoshimura M. Design Sensitivity Analysis of Frequency Response in Machine Structures [J]. Journal of Mechanical Design, 1984, 106 (1): 119–125.

[168] Ong J H. Improved automatic masters for eigenvalue economization [M]. Elsevier Science Publishers B. V. 1987.

[169] Matta K. Selection of degrees of freedom for dynamic analysis [J]. Journal of Pressure Vessel Technology, 1984, 109 (1): 114.

[170] Luo Hong, Li Jun, Cao Youqiang, Zhou Kai. Methods for selecting the master and slave degrees of freedom in dynamic condensation technique of finite element models [J]. Journal of Machine Design, 2010, 27 (12): 11–14.

[171] Liu Xiaobao, Du Ping'an, Qiao Xueyuan. Study on effect of master DOF on errors of substructure static condensation modal analysis [J]. China Mechanical Engineering, 2011 (3): 274–277.

[172] Bao Xuehai, Chi Maoru, Yang Fei. Research on the selection method of master degree of freedom in substructure analysis [J]. Machinary, 2009, 36 (4): 18–20.

[173] Guyan R J. Reduction of stiffness and mass matrices [J]. Aiaa Journal, 1965, 3 (2): 380–380.

[174] O'Callahan J C. A Procedure for an Improved Reduced System (IRS) Model [C]. International Modal Analysis Conference. 1989.

[175] Gordis J H. An analysis of the Improved Reduced System (IRS) model reduction procedure [C]. International Modal Analysis Conference. 10th International Modal Analysis Conference, 1992: 471–479.

[176] Wilson E L, Yuan M, Dickens J M. Dynamic analysis by direct superposition of Ritz vector [J]. Earthquake Engineering & Structural Dynamics, 1982, 10 (6): 813–821.

[177] Huang Mingkai, Ni Zhenhua, Xie zhuangning. Application of Ritz vector direct superposition method in wind-induced response analysis of dome roof [C]. National Conference on structural wind engineering. 2004.

[178] Leung Y T. An accurate method of dynamic condensation in structural analysis [J]. International Journal for Numerical Methods in Engineering, 1978, 12 (11): 1705–1715.

[179] Yang Qiuwei, Liu Jike. An improved method for structural finite element modal reduction [J]. Mechanics in Engineering, 2006, 28 (2): 71–72.

[180] Liu Jike, Yang Qiuwei, Zou tiefang. On the model reduction techniques in structural damage identification [J]. Acta Scientiarum Naturalium Universitatis Sunyatseni, 2006, 45 (1): 1–4.

[181] Wang Wenliang. Structural vibration and dynamic substructure method [M]. Fudan University Press, 1985.

[182] Ren Huili, Wang Xuelin, Hu Yujin, et al. Modal truncation method and its application in vibration analysis of complex structures [J]. China Mechanical Engineering, 2008, 19 (8): 889–892.

[183] Rixen D, Farhat C, Géradin M. A two-step, two-field hybrid method for the static and dynamic analysis of substructure problems with conforming and non-conforming interfaces [J]. Computer Methods in Applied Mechanics & Engineering, 1998, 154 (3–4): 229–264.

[184] Chu M. A continuous approximation to the generalized Schur decomposition [J]. Linear Algebra & Its Applications, 1986, 78 (6): 119–132.

Index

ABOUT THE AUTHOR

Mao Huping is a doctor and Professor of the North University of China. He is a Member of the Communist Party of China and a master tutor. In 2011, he graduated from the School of Mechanical Science and Engineering, Huazhong University of Science and Technology, with a doctorate in engineering. Now he is the executive director of the Shanxi Vibration Engineering Society. His research direction is engineering structure analysis and optimization design.

He has presided over one general project of the National Natural Science Foundation of China, one general project of Shanxi Natural Science Foundation of China, and one horizontal project of enterprises. As a major member, he has participated in one project of National "863" High-tech Research and Development Program, two projects of National Natural Science Foundation, two projects of General Equipment Group, one project of Shanxi Province Joint Fund, and one project of Air China Development Institute. He has written two monographs, one invention patent, participated in compiling one textbook, published 63 academic papers, and included 10 papers in EI.